**ENVIRONMENTAL**
**A**nalyze **C**onsider options **T**ake action

# HABITAT
# and Biodiversity

# ENVIRONMENTAL ACTION
Analyze Consider options Take action In Our Neighborhoods

# HABITAT and Biodiversity

A Student Audit of Resource Use

TEACHER RESOURCE GUIDE

E2: ENVIRONMENT & EDUCATION

DALE SEYMOUR PUBLICATIONS®
ORANGEBURG, NEW YORK

Developed by E2: Environment & Education™, an activity of the Tides Center.

Managing Editor: Cathy Anderson
Senior Editor: Jeri Hayes
Production/Manufacturing Director: Janet Yearian
Design Manager: Jeff Kelly
Senior Production Coordinator: Alan Noyes
Text and Cover Design: Lynda Banks Design
Art: Rachel Gage, Andrea Reider
Composition: Andrea Reider
Clip Art Illustrations: Copyright © Art Parts, Courtesy Art Parts, 714-834-9166

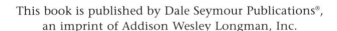

This book is published by Dale Seymour Publications®,
an imprint of Addison Wesley Longman, Inc.

Dale Seymour Publications®
125 Greenbush Road South
Orangeburg, NY 10962
Customer Service: 800 872-1100

Copyright © 1998 The Tides Center/E2: Environment & Education
All rights reserved.
Printed in the United States of America.

Limited reproduction permission: The publisher grants permission to individual
teachers who have purchased this book to reproduce the blackline masters
as needed for use with their own students. Reproduction for an
entire school or school district or for commercial use is prohibited.

Printed on acid-free,
85% recycled paper
(15% post-consumer),
using soy-based ink.

ISBN 0-201-49532-5
DS36860
1 2 3 4 5 6 7 8 9 10-ML-01 00 99 98 97

# ACKNOWLEDGMENTS

Tom and Rampa Hormel, Christopher and Luanne Hormel, Angela Hormel Ocone, and Diane Ives from the Global Environment Project Institute (GEPI) for their unwavering support in developing this action program for students

Jay Dean Paschall of Global Learning and Observation to Benefit the Environment (GLOBE) for developing the concept for this program (originally called EarthTime)

Human-i-Tees for generous support of the review and pilot school portion of the program

The Center for Environmental Education for use of their resource materials

Leslie Crawford and Amy E. Holm for writing and producing *Environmental ACTION*

## EarthTime Contributors

Larry J. Barnes, Ketchum, ID
Betty Bell, Ketchum, ID
J. B. Burrell, Ketchum, ID
Scott Graves, T.R.E.E., Boise, ID
Irene S. Healy, Wood River High School, Hailey, ID
Carrie Hislaire, Sun Valley, ID
Michelle Richer, Ketchum, ID
Doug Wilson, Ketchum, ID

## Environmental ACTION Contributors

Martin J. Byhower, Chadwick School, Palos Verdes, CA
Jennifer Daza, E2: Environment & Education
Bruce Harlan, St. Matthew's Parish School, Pacific Palisades, CA
Anne Kirstin Holm, student at Yale University
Laura Jean Moore, San Diego, CA
Bonnie Slagel, Los Angeles, CA

## Pilot School Participants

Michele Cheyne, Hamilton High School, Milwaukee, WI
Neil Coen, East Magnet School, Kansas City, MO
Steve Engleman, Paul Revere Middle School, Los Angeles, CA
Leroy Hickerson, Alcee Fortier Senior High School, New Orleans, LA
Stephen Hopkins, Sly Park Environmental Education Center, Pollack Pines, CA
Dick Jordan, Boise High School, Boise, ID
Donna Gross McDaniel, Alcee Fortier Senior High School, New Orleans, LA
Dennis Pilien, Bravo Medical Magnet, Los Angeles, CA
Scott Sala, Cory Elementary School, Denver, CO
Pat Shepard, Miller-South School for Visual and Performing Arts, Akron, OH

## Educational Reviewers

Karl Abrahms, Saddleback College, Mission Viejo, CA
Tammy Bird, Crenshaw High School, Los Angeles, CA
Alison L. Brown, Beekmantown Central School, Plattsburgh, NY
Martin Byhower, Chadwick School, Palos Verdes, CA
Joan Grimm, Department of Environmental Quality, Portland, OR
Bruce Harlan, St. Matthew's Parish School, Pacific Palisades, CA
Kay Ice, Educational Development Specialists, Lakewood, CA
Kurt Leuschner, Mira Costa High School, Manhattan Beach, CA
Stephanie Wald, Allan Hancock College, Santa Maria, CA

## Professional Reviewers

Christopher Balthasar, EarthSave Foundation, Santa Monica, CA

Mari Clements, Nutrition Information Center, Crozer-Chester Medical Center, Upland, PA

Wayne R. Gould, Southern California Edison, Rosemead, CA

Joe Haworth, Sanitation Districts of Los Angeles County, Whittier, CA

Richard Heede, Rocky Mountain Institute, Snowmass, CO

Steven Hulbert, Saturn of Olympia, Olympia, WA

Sherman Morrison, North American Coalition on Religion and Ecology, Washington, DC

Gary Petersen, Recycle America, Los Angeles, CA

## E2 Board of Advisors

Peter Corcoran, Bates College

Irene Healy, Wood River High School

Christopher Hormel, Global Environment Project Institute (GEPI)

Rampa Hormel, GEPI

Tom Hormel, GEPI

Dean Paschall, Global Learning and Observation to Benefit the Environment (GLOBE)

Drummond Pike, Tides Foundation and Tides Center

# CONTENTS

## Welcome to Environmental ACTION!

| | |
|---|---|
| About the Environmental ACTION Program | 3 |
|    Program Mission and Objectives | 3 |
|    Program Description | 4 |
|    Program Components | 5 |
|    Moving Through a Module | 6 |
|    Teaching Environmental ACTION | 8 |
|    Community Involvement | 9 |
|    Cooperative Learning | 10 |
|    What's Next? | 13 |
|    Environmental ACTION Feedback | 13 |
| About This Module | 14 |
|    Teaching *Habitat and Biodiversity* | 14 |
|    Assessment Tools | 16 |
|    Cross-Curricular Suggestions | 17 |

## EXPLORE the Issues

| | |
|---|---|
| **1** Recognize Biodiversity | 23 |
|    Activity Sheet 1: Importance of Biodiversity | 25 |
| **2** Recognize Biomes | 27 |
|    Activity Sheet 2: Describe Your Bioregion | 30 |
| **3** Identify Threats to Biodiversity | 32 |
|    Activity Sheet 3: Actions that Impact Biodiversity | 34 |
| **4** Maintain and Increase Biodiversity | 36 |
|    Activity Sheet 4: Design for Biodiversity | 38 |

## ANALYZE

| | | |
|---|---|---|
| **5** | Learn About Community Habitats | 41 |
| | Activity Sheet 5: Notes on Community Habitats and Biodiversity | 43 |
| **6** | Tour the School Campus | 45 |
| | Activity Sheet 6: Campus Landscape Features | 47 |
| **7** | Prepare Your Audit | 48 |
| | Activity Sheet 7: Campus Audit Plan | 49 |
| **8** | Conduct Your Audit | 51 |
| | Activity Sheet 8: Details of Study Area | 53 |
| **9** | Research Plant Species | 55 |
| | Activity Sheet 9: Data Sheet for Landscape Plants | 56 |
| **10** | Summarize Findings | 57 |
| | Activity Sheet 10: Campus Habitats | 59 |

### Act Locally — 61

## CONSIDER OPTIONS

| | | |
|---|---|---|
| **11** | Brainstorm Landscaping Ideas | 65 |
| | Activity Sheet 11: Landscaping Options | 67 |
| **12** | Weigh the Costs and Benefits | 68 |
| | Activity Sheet 12: Assess Costs and Benefits | 70 |
| **13** | Make Recommendations | 71 |
| | Activity Sheet 13: Landscaping Proposal | 72 |

### Act Locally — 73

## TAKE ACTION

 Choose Landscaping Measures — 77

    Activity Sheet 14: Rating Sheet — 79

 Prepare and Present Your Proposal — 80

    Activity Sheet 15: Proposal Checklist — 82

 Track Response to Proposal — 84

    Activity Sheet 16: Tracking Sheet — 86

## Appendices

| | |
|---|---|
| **Issues and Information** | 91 |
| Section A  Biodiversity | 91 |
| Section B  Biomes, Bioregions, and Habitats | 93 |
| Section C  Threats to Biodiversity | 95 |
| Section D  Conserving Biodiversity | 97 |
| Section E  Plants | 100 |
| Section F  Symbols to Use in Drawings of Landscape Plans | 102 |
| Section G  Principles of Sustainable Gardening and Landscaping | 103 |
| Section H  Tips for Planning and Maintaining a Sustainable Garden | 104 |
| Section I  Garden and Landscape Design | 107 |
| Section J  Organic Gardening | 110 |
| Glossary | 112 |

| | |
|---|---|
| **Teacher Resources** | 113 |
| Organizations | 115 |
| Government Agencies | 118 |
| Books and Pamphlets | 119 |
| Products and Services | 122 |

# Blackline Masters

| | |
|---|---|
| **Activity Sheets** | 127 |
| Assessment Tools | |
|     Content Quiz | 152 |
|     Student Survey | 154 |
|     Student Self-Evaluation Form | 155 |
|     Action Group Evaluation Form | 156 |
|     Program Evaluation Form | 157 |

# Welcome to Environmental ACTION!

# About the Environmental ACTION Program

As the natural resources crisis reaches global proportions—air, water, and land pollution; limited supplies of fossil fuels; threatened and endangered habitats; species facing extinction—people must acknowledge the enormous role that they have played in creating or exacerbating the problems and the enormous role they can potentially play to alter the course of events. As educators, we feel it is vital to give students the information and skills they will need in order to take action. We need to help students learn to look critically at environmental issues and take personal responsibility for finding solutions by asking questions: What is the problem? What causes the problem? What impact does behavior have on the problem? What changes can be made or actions taken to help? By introducing an action-based curriculum, we not only begin preparing the next generation for dealing with difficult issues, we teach students how to live healthier lives in the process.

The school is an ideal laboratory for this hands-on experiment because it provides a real-world model. Students gather information about the school, analyze environmental issues within the context of the information gathered, determine positive alternatives, and practice implementing solutions right in the school setting. For example, students may decide that by using energy-efficient lighting, implementing water conservation programs, moving toward environment-friendly landscaping, or using nontoxic chemicals to clean and maintain facilities, they can improve the school environment, reduce resource use, and perhaps even save money. They present their proposal for changes to the school for approval. If the changes are implemented, the school then becomes a useful living paradigm; students can participate in and observe the changes, monitor results, and extend their knowledge to their own homes. It is this learning/doing, school/community partnership that makes Environmental ACTION a unique environmental education curriculum.

## Program Mission and Objectives

Environmental ACTION was developed in response to the need for environmental education materials that emphasize personal responsibility and positive action. The program's mission is to empower students with the knowledge and skills necessary to make meaningful changes that can be carried into the future. Many educators have been challenged to find appropriate materials and the necessary teaching support to offer this kind of environmental education to their students. Environmental ACTION meets these needs by providing a relevant, supportive, clearly structured curriculum specifically aimed at middle and secondary students.

The six modules in the program provide step-by-step instruction on how to investigate real-world environmental issues and present opportunities to learn and practice action skills in the context of these issues. By creating a laboratory within the school community, the program gives students the opportunity to learn and develop personal responsibility through practical application.

The curriculum has the following objectives:

- to promote awareness of environmental issues through real-world investigations
- to build the knowledge and skills needed to analyze, investigate, and offer solutions to environmental problems

- to encourage practical application of knowledge and skills in issue resolution
- to assist students in becoming responsible citizens by involving them in the extended school community

## Program Description

The program consists of six modules designed for use in middle and secondary schools. Each module includes a Teacher Resource Guide and Student Edition. Following is a brief description of the six modules:

### Energy Conservation
Students explore the sources, production, uses, and environmental effects of energy. They apply their learning by examining ways to improve the energy efficiency of their school and homes.

### Food Choices
Students investigate the effects of food production, diet, and nutrition on human health and the environment. Students analyze the school's food service program to identify healthy choices and practices.

### Habitat and Biodiversity
Students study the importance of biological diversity, landscape management, xeriscaping, composting, and integrated pest management (IPM). Using the school as a research laboratory, students assess the current landscaping, then evaluate its present health and environmental impact. This module also contains a step-by-step guide to creating an organic garden and seed bank.

### Chemicals: Choosing Wisely
Students investigate the types of materials, chemical products, cleaning supplies, and pesticides used in their school—how they are used, stored, and disposed of, and their potential effects on human health and the environment. Students develop a plan for implementing Earth- and human-friendly alternatives for school and home.

### Waste Reduction
Students sort and analyze school garbage to identify recyclable and compostable materials. They formulate a plan to reduce their consumption and waste at school and at home, including developing a recycling program or improving an existing one.

### Water Conservation
After an introduction to water consumption and quality issues, students conduct an audit of water usage and efficiency on the school campus. Using the school as a research laboratory, students develop strategies for implementing water conservation at school and at home.

## Options for Using the Program

These six modules are part of a complete yet flexible curriculum package. Because they are designed to be used either in conjunction with one another or as stand-alone units, a variety of teaching options are offered. For example, one module can be taught year after year, different modules can be taught in consecutive years, or different modules can be taught in a given year. Modules may be taught by a team, or a group of teachers may choose to simultaneously present different modules.

## Using the Program Outside School

In response to a growing number of requests, Environmental ACTION has been designed for use by groups that operate outside a formal school setting. Issue investigation and action activities in the Teacher Resource Guide and Student Edition can be easily adapted to fit different venues and circumstances. For example, an ecology club might use the program at a school, community center, or local business; church groups can use it to explore their stewardship responsibilities within the church community or as an outreach project; community-based organizations can conduct the activities at their centers as part of their efforts to improve the quality of life within the local community. Wherever choices are made with respect to natural resource consumption, there exists an opportunity for investigation, evaluation, and action—and Environmental ACTION can provide a framework for these efforts.

## Program Components

Each module consists of a Student Edition and a Teacher Resource Guide that complement one another and provide maximum flexibility for teachers.

## Student Edition

Each activity is presented using a clear and consistent format that puts students in charge of their own learning. Setting the Stage offers discussion questions that will pique students' interest and direct their attention to the objectives of the lesson. Vocabulary is included as appropriate. The Focus part of the activity provides the core of the lesson, leading students through a step-by-step process. Central to this section is an activity sheet, which can be duplicated for students to complete individually or in groups. (Blackline masters for all activity sheets are located at the back of the Teacher Resource Guide.) It's a Wrap reviews student work and provides closure to the activity. The Home activity allows students to apply their learning outside the school environment.

## Teacher Resource Guide

The material in the Teacher Resource Guide duplicates key text in the Student Edition—questions, assignments, and activity sheets. Concepts and objectives are itemized, special vocabulary is listed, and student responses are suggested. Activity sheets in the Teacher Resource Guide include annotated answers. The Teacher Resource Guide contains the same Issues and Information section and Glossary that appear in the Student Edition, along with some additional resources and assessment tools. The blackline masters for all activity sheets are located at the end of the Teacher Resource Guide.

## Moving Through a Module

Each activity within a module is a separate lesson, and most are designed to take only one class period. The program consists of four types of activities: Explore the Issues, Analyze, Consider Options, and Take Action. The teaching methodology, illustrated below, presents activities in a sequence that first introduces students to global environmental issues, then requires them to apply critical thinking skills to environmental issues in their immediate environment, and finally encourages them to take responsibility through independent action.

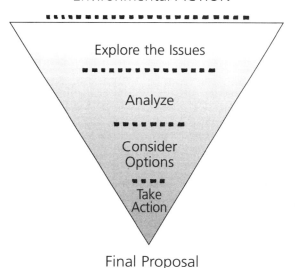

### Explore the Issues Activities

In the first four activities in each module, students are given an overview of the environmental topics that pertain to the issues they will focus on in their investigation of school, home, and community. The Explore the Issues activities gives students opportunities to gather information and build environmental awareness. Students will work individually, in small groups, and as a class.

### ACT Activities

The remaining activities are structured to provide an environment for students to move from awareness to choice to action. The three-part teaching sequence—ACT—reflects the following process:
- Analyze
- Consider Options
- Take Action

**Analyze Activities** Students focus their attention on the school campus or other facility. Working in small groups, students develop and implement a plan for collecting and recording data. Finally, students identify environmental, financial, and human health costs and benefits of current school practices affecting, for example, energy use, waste disposal, or whatever topic is the focus of the particular module.

**Consider Options Activities** Students investigate alternative practices, services, technologies, and products. Then they evaluate each on a costs/benefits basis, using material in the Issues and Information section at the back of the book.

**Take Action Activities** Based on the options that students have discovered, groups of students recommend changes in practices and products. The class then discusses and evaluates the recommendations and ultimately reaches a consensus as to which recommendations to propose to the school administration or environmental committee. (Note: See Community Recognition, p. 9, for further discussion on local outreach.) The closing activities involve writing the proposal. This final step in each Environmental ACTION module is intended to launch students on a course for creating a healthier and more balanced environment.

### Home

Each activity includes ideas for parallel investigations and analyses that students can do at home. These activities reinforce the learning going on at school and provide an opportunity for students to apply the knowledge to practices and choices made at home. Family members may also choose to get involved. The Home activities are optional; they are not required for the completion of the school environmental audit or other activities within the module. We suggest that students' Home assignments be written in a Journal. (See Student Journals, below.)

## Act Locally

In order to effect meaningful change in areas related to human health and the environment, it is essential that students practice taking action. As with other areas of learning, practicing a behavior makes it grow into habit. Therefore, Act Locally activities—straightforward outreach projects to present to the school or community—are provided for use after the Analyze and Consider Options sections. Students can choose to implement one of the suggested activities or develop an activity of their own. In either case, these Act Locally activities serve as opportunities for students to practice putting their ideas into action.

## Student Journals

Student work on the Home assignments at the end of each activity can be recorded in student Journals, along with notes, sketches, data collection, and other student writing that takes place during the module. Journals may be kept private and used for referral and note-taking during discussions, or they may be used to record answers to specific questions in the Focus or It's a Wrap sections. Since some Journal material may be turned in and some kept for personal use only, a loose-leaf notebook or folder is an ideal choice.

You may want to extend the use of the Journal by suggesting additional writing assignments that call on students to make observations about what they have noted, synthesize or analyze the data, compare and contrast information, draw conclusions, or make generalizations. All of these activities provide opportunities for students to use critical thinking skills and communicate ideas—valuable experiences for most middle and high school students. Additional ideas for Journal assignments can be found in the Cross-Curricular Activities Chart for the module. (See Cross-Curricular Suggestions, below.)

## Issues and Information

An Issues and Information appendix appears in both the Student Edition and the Teacher Resource Guide. This appendix contains a variety of materials students need to complete the activities and develop a deeper understanding of the issues they are studying. The appendix is divided into easily accessible sections and, in some modules, may be accompanied by illustrations.

## Cross-Curricular Suggestions

At the end of the introductory material for each module, you will find specific suggestions for cross-curricular activities and extensions in the following content areas: history/social studies, language arts, math, and science. These suggestions offer opportunities to extend study activities to higher cognitive skill levels and/or to meet your school's curriculum framework with activities that link directly to specific components.

# Teaching Environmental ACTION

**Scheduling and Managing Modules** Start a module with enough time—18–20 class sessions—for students to conduct research and implement changes. Since students will be needing activity sheets for all the activities, you may wish to photocopy them before starting a module. The blackline masters are located at the end of the Teacher Resource Guide. Before starting the first activity, introduce students to the project, administer the Content Quiz and Student Survey provided in the Blackline Masters/Assessment Tools section, and familiarize students with the Student Edition. You may also wish to have students write to agencies, organizations, manufacturers, and businesses (listed in the Teacher Resources section of the Teacher Resource Guide) in order to obtain materials and request information that can be used as the module progresses. Once the students are working in their Action Groups to investigate and analyze their subjects, additional research time outside of the actual class sessions may need to be scheduled.

**Skipping or Modifying Activities** Since all of the activities have been carefully designed to help students build their investigation of the school environment, we encourage you to complete each activity. If you do omit or modify an activity, be sure that students incorporate the activity's objectives into the project. You may be able to teach or facilitate your students' comprehension of a learning objective in a different way or combine activities in order to make sure that your students do not miss meaningful steps in the discovery process.

**The Teacher's Role** The Environmental ACTION modules are activity driven. The scope and sequence has been designed carefully so that the transition from teacher- to student-directed activities happens gradually and within well-outlined parameters. In the beginning, you will be in the more traditional role of the teacher, providing information and leading the students through the Explore the Issues activities. As the class begins the ACT activities, you will assume the role of facilitator for your students' efforts.

Small group work will be the activity most likely to affect your role in relation to the students. Although you will continue to supervise them, you will not be directing their work. Students working in Action Groups will be responsible for keeping on task, catching their mistakes, supporting each other, and meeting their goals. Students may not always carry out their work in a way that you would have directed, but this learning process allows them to make mistakes and correct their actions.

As your students work through the actual audit and your role becomes more that of a facilitator, your responsibilities may include some or all of the following:

- serving as liaison between the students and the school administration
- keeping the students' work on schedule
- monitoring the groups' interactions
- advising Action Groups about procedures and resources and assisting with problem-solving
- reserving reference and resource books in the school library or classroom for student research

Toward the end of the ACT sequence, students will meet again as a class to finalize their proposal. At this point, you will evaluate the group dynamics of your students and assess the degree of facilitation necessary. After students write the proposal and present it to the school administration or environmental committee, the actual implementation steps will be determined by the school's willingness to take action and your students' commitment to the project.

**Student Project Coordinator** To provide more leadership opportunities, you may wish to create the role of Project Coordinator for one student or several students. The Project Coordinator can volunteer, can be appointed by you, can be selected by the class, or can be chosen at random to assist you with the administration and facilitation of the project in the following ways:

- Lead large group sessions
- Serve as liaison between the Action Groups and you or the class and the school administration
- Help to monitor and keep Action Groups on schedule
- Oversee the preparation and distribution of materials, including activity sheets

## Community Involvement

**Community Participation** The Environmental ACTION modules enable students to participate in community service activities and help students develop a commitment to public service. For optimum success, you and your students will want to cultivate the goodwill of all potential participants from the beginning of the project. These participants may include other students, teachers, administrators, librarians, kitchen staff, custodians, maintenance staff, and purchasing personnel, as well as members of the community. The following are some hints that may be helpful in gaining the participation and cooperation of members of your community:

- Notify key support groups in your school and community of your project.
- Provide all participants with a description of the activities and a description of things students will be doing in the course of their work.
- Make a formal request for any specific considerations or resources needed from these individuals.
- Inform students' parents. They may be able to contribute additional resources and/or expertise. Also, parents' understanding of homework assignments such as the home audits will help ensure a positive and productive experience.
- Recognize the cultural, economic, and ethnic diversity of your school community and be sensitive to the varied perceptions of issues related to human health and the health of the environment.

**Community Recognition** It is important for the students to realize that their responsible investigations of an environmental issue and development of a logical plan for taking action have earned serious consideration by the community. Therefore, it is critical to engage community involvement that will include a thoughtful response to the students' final report and proposal.

One way to encourage community participation is to create an environmental committee at your school. This committee can be set up at the beginning of the project. The members may include parents, teachers, students, administrators, and other school staff. If a school Ecology Club is already in place, you may wish to explore establishing a committee under the auspices of the club. The committee can serve as a resource, a liaison between the students and school officials, and a review body for ideas and suggestions for implementing changes. In order to be effective, the committee will probably need to meet periodically throughout the project to hear status reports and to respond to requests for assistance or advice. You may want to draft a written agreement detailing the role of participants and the service each agrees to provide.

Recognition of students' efforts to bring about positive change in the school setting could include a public award ceremony, certificates of merit, academic or extra-curricular credit, or establishment of an Implementation Committee on which students might serve.

**Bringing Role Models into the Classroom**
The first activity in each Analyze section suggests that you ask leaders from your community or resource people who are familiar with the school to present information related to the module issue. Plan ahead to find and make arrangements for guest speakers. It can be a time-consuming task, but outside participation in this project is important for a number of reasons:

- The information provided will help students understand how the issue relates to their community, school, family, and themselves.
- Students will be able to ask questions and obtain answers that will assist them in their investigation and audit of the school campus.
- Role models from your community can become valuable and accessible resources for you and your students.

What kinds of role models make effective guest speakers? The main criterion is that they are familiar with the module issue. You may want to invite a school administrator, parent, elected official, resource person with a government agency or advocacy organization, business owner, medical professional, or a family member of one of your students. Your students may have some good ideas about whom they would like to meet and hear speak. You might invite a number of guests to allow students to hear several points of view.

A fun and interactive way to present a speaker is to introduce the guest to your students without revealing the person's connection to the module issue. Then play Twenty Questions, having students ask yes/no questions to try to discover the person's role. Once the role is determined or the maximum number of questions has been asked, you can introduce the guest and provide some background.

## Cooperative Learning

Cooperative learning—including small-group discussion, problem-solving, and decision-making activities—is an integral part of the Environmental ACTION curriculum. Because some students may not have previous experience with these processes or may not yet have developed strong skills, you may need to prepare them for this type of interaction. The effectiveness of small group discussion as a learning tool varies widely according to several fairly concrete parameters, and you can influence the outcome in your classroom by sharing these with your students, thus providing a model for them to work toward.

First, we believe that small group discussion is most effective when the number of participants is limited (four to six) to allow members to be aware of and react to one another. Additionally, it is critical that participants have a common understanding of goal(s), have a sense of being part of the group and are willing to work cooperatively, work toward open dialogue and discussion among participants, and recognize that individual accomplishment depends upon the group's success.

Following are some questions and answers about basic cooperative learning terminology and ways to manage small group experiences. Many of these topics are discussed in more detail in *Effective Group Discussion* by John K. Brilhart (Wm. C. Brown Company Publishers, Dubuque, Iowa).

Q: What makes an effective Action Group member?

A: Characteristics include a sense of responsibility for helping the group accomplish its objectives, active participation, willingness to accept tasks, and commitment to the group through good and bad. Also, effective group members offer positive feedback on ideas and information, avoid using words that induce negative feelings, make organized comments, speak concisely, state one point at a time, pose specific questions, and listen courteously and attentively. It is to be expected that students will improve their skills in these areas as they gain experience working in their Action Group.

Roles for individual group members will evolve through the workings of the group. In some cases, there may be no apparent leader but lots of leadership. Conversely, groups may have leaders but lack direction. The best-case scenario is for each group to have a leader who oversees procedures and facilitates the group's process.

Q: How should Action Groups be organized physically?

A: Action Groups work best when they are in comfortable environments that do not offer a lot of distractions. Students should sit close together in a circle and have access to writing surfaces. If you notice some awkwardness when students are first assigned to Action Groups, you might want to introduce an ice-breaker activity to help them get comfortable with working together.

Q: What can I expect from students working in Action Groups?

A: Small groups usually progress through several steps before achieving end goals. Groups vary in terms of the time it takes to move through some or all of the following steps:

- Members of the group work to develop a cohesiveness of goals and understanding of one another.
- The group establishes the identities and roles of individual members, including leadership roles. This step may be accompanied by disagreements.
- The group finds ways to manage disagreements by establishing rules and procedural standards. This usually results in more agreement and an increased feeling of freedom to offer opinions and have them heard.
- The group focuses on the goal(s), and discussion is concentrated on developing solutions to the problems identified.

Q: Is decision-making the same as problem-solving?

A: No. Problem-solving involves a number of phases—identifying the problem; investigating alternatives; developing an implementation plan; and making the necessary changes. Problem-solving sessions should focus on the issue and a careful framing of the problem question before discussing possible solutions. They should also encourage participants to share their knowledge about the issue and refrain from making judgments when discussing alternatives. Decision-making comes into the problem-solving process whenever students choose between various options or possible courses of action.

Q: When is group brainstorming an appropriate process?

A: Brainstorming is an appropriate small group discussion tool when a number of solutions exist for a particular problem. The following conditions of brainstorming should be communicated to participants before brainstorming sessions begin:

- Any idea is valid; the more, the merrier.
- Ideas can build upon previous suggestions.
- Suggestions should not be judged until brainstorming has ended.

Q: What is consensus decision-making?
A: Consensus decision-making is an alternative to the more traditional majority vote group decision-making. In the consensus process, value is placed on each member of the group agreeing to a decision. It may not be the preferred decision of any individual, but it is something everyone can live with. This type of decision-making allows for the group to meet a multitude of criteria for making a choice and for each member to derive satisfaction from the final result. However, consensus decision-making can take considerably longer than other processes such as voting, and in some cases it is not feasible to achieve total agreement.

If you choose to make use of consensus decision-making, you may want to follow certain steps. First, list options, stating reasons in favor of and against them. Next, rank options according to criteria such as accessibility or time constraints. Then, consolidate ideas and summarize the class discussion. Next, narrow down the options to one or two. Using the chalkboard, you might list under each option the pros and cons. Ask for a class vote. If the vote is not unanimous, ask volunteers for each side to explain their reasons. Encourage students to use persuasive arguments but also to listen attentively to the opposing arguments. Continue polling the class until finally all are in agreement about their choice of a study area. Activity sheets are provided for this purpose.

Q: If an Action Group is not staying focused, what can I do?
A: You can help by asking one or more of the following questions:

What are you meeting about? What is your common goal? What do you want to achieve as a group? Are you staying on track with your tasks?

## Evaluating Students' Work

There are numerous ways to evaluate students' work and progress. We suggest you begin with the Student Survey provided in the Blackline Masters/Assessment Tools section at the end of this book. This survey contains both objective and subjective questions and can be given at the beginning and the end of the project. If used at the beginning, the process of reviewing the students' answers can serve as a device for discussion or as a tool for assessing your students' base level of knowledge about a particular issue. If used again at the end of the project, the survey can help you evaluate the amount of learning and changed attitudes/behaviors that resulted from students' investigations and activities.

To evaluate different aspects of an individual student's participation in the Action Groups, you may choose to use the Action Group Evaluation Form. This is a simple checklist for your reference. The blackline master is provided in the Assessment Tools section on page 156 of this book. Another evaluation tool that works well in a cooperative setting is a self-evaluation submitted by students. A blackline master Student Self-Evaluation Sheet is provided on page 155 of the Assessment Tools section. In order to best assess long-term performance, you will want to schedule evaluations at regular intervals throughout the project.

## What's Next?

Teachers often wonder what happens after their class completes a module and initiates changes or environmental activities at the school. What is next semester's or next year's class supposed to do? As long as the school has areas that need improvement, learning opportunities abound. The following are some options you may find useful:

- Repeat the module activities to evaluate the impact of changes, and continue to develop additional recommendations including modifications of the previous changes.
- Maintain the newly instituted activities at the school.
- Choose another module, and allow your students to explore a different environmental issue.

## Environmental ACTION Feedback

We are interested in the ongoing short- and long-term effects of implementing our curriculum at schools, so we invite you to provide feedback about your experience. The data collected will not only provide us with information about the effectiveness of the curriculum, it will allow us to compile information illustrating the degree of change occurring on campuses nationally and abroad. This information also becomes a peer-networking resource for teachers and students who have questions about implementing environmental programs at their schools. We ask that you take a few minutes at the conclusion of this project to complete the Program Evaluation Form located at the end of this guide. We also urge you, at any time, to visit our Web site. The address is: **<http://www.enviroaction.org>**. Let's share our successes and challenges! We'll add your contribution to our site.

# About This Module

## Teaching Habitat and Biodiversity

Begin by giving students an overview of the module and its activities. Tell them that this project is aimed at enhancing habitats and maintaining or increasing biodiversity on the school campus. Explain that they will be studying different campus sites to audit use by both people and wildlife so that they can develop a plan for landscaping practices aimed at increasing biodiversity.

Distribute the Student Editions and ask your students to read the Welcome! section. Allow time for discussion of the students books, teacher/student roles, evaluation, schedules, contact with the school administration, procedures, and Student Journals.

To assess students' prior knowledge of this topic before they begin work and to get them thinking critically about habitat and biodiversity, you may want to have students complete the Student Survey provided in the Blackline Masters/Assessment Tools section of this guide, page 154. A discussion about using the survey as an assessment tool appears on page 16.

Since much of the work in this program will be done in small groups, you may want to take time at this point to explain the methodology of Action Groups and to offer tips on how to work cooperatively. (See pages 10–12 for specific suggestions on facilitating Action Groups.)

**Time Requirements** Most activities are designed to be conducted in a single 50- or 55-minute class session. The first and second Analyze activities may take more than one session, depending on how many guests are invited to speak and how the tours are organized. Also, Action Group work and homework in several of the activities require extra time outside of class. The amount of time assigned to homework and independent group work can vary.

**Materials and Preparation** Implementation of the *Habitat and Biodiversity* module requires no special materials or equipment. Garden books and field guides for birds in your area would be helpful to have on hand. Students will need pencils and paper to take notes and make Journal entries while surveying the school. Poster board, oak tag, construction paper, markers, and general art supplies will help students make effective presentations.

Blackline masters for the activity sheets are located at the end of this book. You may choose to photocopy all of them before beginning a new module, or one at a time at the appropriate point in the lesson. See the Prepare section of each activity for any additional preparation required.

**Explore the Issues Activities**

The Explore activities emphasize "the big picture," presenting data for students to analyze and use to create diagrams, charts, and graphs, and offering ideas about how responsible individual and/or group action can effect positive environmental change. Students become aware of different habitats and how they are threatened worldwide. The activities give students experience working independently, in small groups, and as a class.

**Activity 1: Recognize Biodiversity** Students learn about biodiversity and the interdependence of all living things on earth. They read a passage to discover ways in which changes to one species can affect the survival of another.

**Activity 2: Recognize Biomes** Students find out how living things depend on the environment to meet their needs, as well as features of the environment in different parts of the world.

**Activity 3: Identify Threats to Biodiversity** Students discover how biodiversity is threatened and what can be done to maintain or increase biodiversity.

**Activity 4: Maintain and Increase Biodiversity** Students find out how they can make a difference in maintaining and increasing biodiversity where they live.

**ACT Activities**

Once students have an overview of biodiversity and factors that affect it, they embark on the ACT activities. These activities guide students through the process of conducting an audit of the school habitats (Analyze activities); investigating ways to enhance habitats and maintain or increase biodiversity (Consider Options activities); and proposing changes in landscaping practices (Take Action activities).

**Analyze Activities**

**Activity 5: Learn About Community Habitats** Students learn about local habitat and biodiversity issues from a community resource person.

**Activity 6: Tour the School Campus** Students learn about habitat and biodiversity issues on the school campus from a resource person. They tour the campus and draw a rough sketch of the landscape features.

**Activity 7: Prepare Your Audit** Using the sketch from Activity Sheet 6 and their campus tour notes, students will divide the campus into study areas for the audit.

**Activity 8: Conduct Your Audit** Action Groups will inspect their study areas and use their observations to create detailed diagrams.

**Activity 9: Research Plant Species** Students will research plant species to find out more about their study areas and how they are maintained.

**Activity 10: Summarize Findings** Students will review audit results, evaluate how habitats are maintained, and describe how biodiversity is supported and write a summary of their findings.

**Consider Options Activities**

**Activity 11: Brainstorm Landscaping Ideas** Students decide how they can protect or enhance habitats found on the school campus and investigate landscaping practices that will maintain or increase the biodiversity in those habitats.

**Activity 12: Weigh the Costs and Benefits** Students use the information they have gathered to evaluate the costs and benefits of the landscaping practices they are proposing to adopt.

**Activity 13: Make Recommendations** Students working in Action Groups give final consideration to all the landscaping options they have been exploring. After weighing the costs and benefits of each, they will select the best ones and develop a finished proposal for presenting the recommendations to the class.

**Take Action Activities**

**Activity 14: Choose Landscaping Measures** Action Groups will present their recommendations to the class. The class will discuss the various suggestions and reach a consensus about those recommendations they will include in the final proposal to be presented to school officials.

**Activity 15: Prepare and Present Your Proposal** Students will write a proposal for implementing water conservation measures at school. They will outline costs, benefits, and describe the ways to implement their recommendations. Then they will present it to the school committee.

**Activity 16: Track Response to Proposal**

Students will follow up on their landscaping proposal as it is implemented on campus. They will work to increase awareness of the importance of enhancing habitats in order to maintain or increase biodiversity.

## Assessment Tools

The following assessment tools have been provided in the Blackline Master/Assessment Tools section of this guide: Content Quiz, Student Survey, Student Self-Evaluation Form, and Action Group Evaluation Form. At this time you may wish to administer the Content Quiz and the Student Survey as pre-tests for the module. These two tools are intended to be used again, as post-tests, at the end of the module. A reminder has been included on page 81 of this guide. Answers to the Content Quiz are shown below. The Student Survey (see page 154) is composed of subjective questions, so answers will vary. The Student Self-Evaluation Form and the Action Group Evaluation Form can also be administered at the end of the module. Finally a Program Evaluation Form has been provided for your optional feedback.

## Answers to Content Quiz

1. b. biodiversity refers to the huge variety of living things on earth
2. a. true (Biodiversity is a threatened resource.)
3. c. near the equator
4. a. true (Living and nonliving things in an ecosystem interact.)
5. d. All of the above
6. a. true (Threats to biodiversity upset the balance of life.)
7. a. true (An ecosystem can depend on one keystone species.)
8. b. false (Biomes with harsh climates and short growing seasons have lower levels of biodiversity.)
9. d. All of the above
10. a. true (Habitat destruction is the greatest threat to biodiversity.)
11. b. destroy the forest habitat
12. d. All of the above

## Cross-Curricular Suggestions

| Description of Activity | Curriculum Connections | When to Use |
|---|---|---|
| Have students interview grandparents and older neighbors and friends about how the landscape of the area has changed over the years. Have students ask how the town or city developed, where roads were built, what was once forest or farmland, and what activities they enjoyed in the countryside. Have students determine the time period in which their subjects were growing up and whether it was primarily a rural or urban area. Have students record their interview notes in their Journals. In a class discussion, ask students to share their findings. Guide the class in comparing the look of the land today with what it once was. | History/Social Studies, Language Arts | Following Activity 2 |
| Have students do research to learn about public parks and gardens in their community or region. You might have them gather information by writing letters or telephoning the local parks department, environmental groups concerned about preserving habitats, your state department of natural resources, the historical society, or local garden clubs. Have them create a pie graph showing the percentage of space in the city limits devoted to parks and public gardens. In class discussion, invite students to suggest ways in which the amount of space devoted to parks and public gardens can be increased. | Social Studies, Mathematics | Following Activity 3 |

Welcome to Environmental ACTION!

| Description of Activity | Curriculum Connections | When to Use |
|---|---|---|
| Have students create a map of their community's waterways: rivers, streams, and lakes. On their map, have them show how the riparian zone is developed or used. Is it a protected habitat? Is it set aside for recreation? Is it accessible? For information, students might turn first to their local water authority. Other information sources might include your state department of natural resources, the U.S. Geological Survey, and any water conservation projects in the area. | Science, Geography/ Social Studies | Following Activity 5 |
| Challenge students to create a field guide to local birds in the area. The guide can include information on behavior, migration, diet, nesting habits, as well as coloring and song. The local Audubon Society can provide information on species that may be of special interest in your area. | Language Arts, Science | Following Activity 11 |
| As an extension of the above activity, students might work together to write an article for their school or local newspaper, featuring birds with campus nesting sites, birds that are threatened or endangered, or a profile of the state bird. | Language Arts, Social Studies, Science | Following Activity 11 |

| Description of Activity | Curriculum Connections | When to Use |
|---|---|---|
| To help students develop a better understanding of how much carbon dioxide is absorbed by trees, have them make the classroom into a visual display. First they can find out how many cars are registered in the city or county or state. Then they can calculate how many acres of trees it would take to clean the air at about one tree per car per year. How many acres of pavement are in the parking lot of the local shopping mall? Encourage students to use their imaginations to make a poster showing what they discover. | Mathematics | Following Activity 13 |

# Explore the Issues

# RECOGNIZE BIODIVERSITY

Students learn about biodiversity and the interdependence of all living things on earth. They read a passage to discover ways in which changes to one species can affect the survival of another.

### Key Concepts

- Living things need each other in order to survive.
- Changes in habitat affect biodiversity.
- Habitat destruction can lead to species extinction.

### Objectives

After completing this activity, students will be able to

- explain the importance of biodiversity
- describe the interdependence of living things
- describe how changes to one species can affect another

### Prepare

You may wish to read about biodiversity in Issues and Information section A.

### Materials

Activity Sheet 1 for each student (See Blackline Masters section.)

### Setting the Stage

Discuss briefly the questions on student page 9.

- **What is biodiversity?**

  Biodiversity refers to the variety of living things on earth.

- **Why is biodiversity important?**

  Biodiversity helps to ensure the health of the planet by maintaining the balance of living things and providing building blocks of evolution. Encourage students to give examples of why they think biodiversity is important.

### Vocabulary

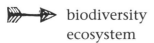
biodiversity
ecosystem
extinction
genetic
habitat
interdependence
species

### Focus

**1.** Have students read the passage on student pages 9–10 and discuss how habitat changes can affect biodiversity in positive and negative ways. Students may also look at the background material in Issues and Information section A. Then have students study the graph that shows species loss during this century and answer the questions. Students should recognize that the rate of extinction has sharply increased in the last decade. The reason for the increase is habitat destruction.

Explore the Issues: Recognize Biodiversity   23

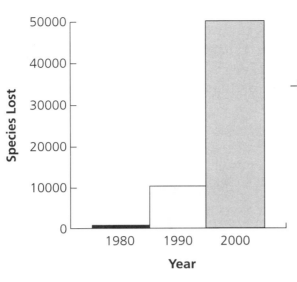

1 species lost in 1900
400 species lost in 1980
10,000 species lost in 1990
−50,000 species lost in 2000

- **What do living things need in order to survive?**

    Responses include clean water, food, clean air, each other.

- **How do living things depend on each other?**

    Responses include providing food sources, protection, balance of predators and prey.

- **What is an extreme result of habitat loss?**

    Students should realize that extinction is the extreme result of habitat loss.

**2.** Encourage students to identify the interdependence of living things in the landscape depicted on Activity Sheet 1.

## It's a Wrap

Once students have identified evidence of interdependence on the skills sheet, have them brainstorm a list of plants and animals and then make their own connections.

## Home

**Assignment** Home activity on student page 11.

 How do you depend on other species of plants and animals? During the next 24 hours, keep track of all the things you and your family use that are connected to another species. For example, furniture comes from trees, clothing comes from plants and animals, food comes form plants and animals. List your connections in your Journal.

ACTIVITY SHEET

Name

# IMPORTANCE OF BIODIVERSITY (part 1)

Study the illustration and answer the questions that follow.

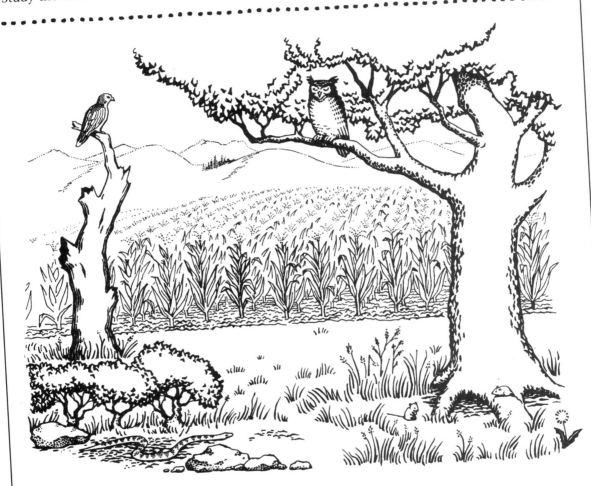

1. What features of this habitat can you identify?
   **meadow, field, trees, agriculture, plants, wildlife**

Explore the Issues: Recognize Biodiversity   25

ACTIVITY SHEET

Name

# IMPORTANCE OF BIODIVERSITY (part 2)

**2.** What could happen to the biodiversity in this habitat if a flood wiped out the corn crop?

**There would be fewer mice, which would affect the snake, owl, and hawk populations.**

**3.** Mice are an important source of food for owls. How could the owls be affected by the loss of the corn crop?

**They would not have mice to eat.**

**4.** How would the owls be affected by the loss of the hawks?

**The owls might have more food to eat because they wouldn't be competing with hawks.**

**5.** What other examples of interdependence can you identify?

**Hawks eat snakes, snakes eat mice, mice eat corn; the birds nest in the trees and snags; the snakes and mice nest in the meadow.**

**6.** What could you do to increase the biodiversity in this habitat?

**Possible answers include plant more trees and shrubs.**

**7.** What could happen if this habitat were developed into a shopping mall?

**All of the animals would need to find other places to live.**

26  Environmental ACTION  Habitat  Teacher Resource Guide

# RECOGNIZE BIOMES

Students find out how living things depend on the environment to meet their needs, as well as features of the environment in different parts of the world.

### Key Concepts

- Species depend on the environment in order to survive.
- Features of each biome influence biodiversity in that area.

### Objectives

After completing this activity students will be able to
- identify what animals need for survival
- describe how living things depend on their environment

### Prepare

You may wish to read Issues and Information section B for background material on the earth's biomes.

### Materials

Activity Sheet 2 for each student

### Setting the Stage

Discuss briefly the questions on student page 13.

- **What do living things need in order to survive?**

  Responses may include water, food, shelter, sunlight.

- **What are some ways in which the basic needs of living things are provided for in your area?**

  Encourage students to describe physical features of your geographical area.

### Vocabulary

 biome
bioregion
chaparral
climate
coniferous
deciduous
precipitation
savanna
taiga
temperate
tropical
tundra

### Focus

**1.** Students study the chart and use it to identify features of the different biomes. Then they relate the features to biodiversity by answering the questions.

- **What biomes have the lowest level of biodiversity?**

  Responses include tundra, taiga, desert.

- **How might a short growing season affect animal life?**

  Responses include a lack of food for long periods of time, migrating animals that impact the environment and then leave.

Explore the Issues: Recognize Biomes  27

| Biome | Climate | Common Plants | Common Animals |
|---|---|---|---|
| Tundra | very cold, dry; deep soil is permanently frozen; long, dark winters | few species; lichens, mosses, low shrubs | wolves, lemmings, polar bears, Arctic foxes, migratory birds; most migrate long distances |
| Taiga | cold winters, short growing season | evergreen shrubs and coniferous trees (fir, spruce, pine); some deciduous trees (birch, aspen) | bears, moose, wolves, ducks, loons, migratory birds |
| Temperate Coniferous Forest | damp, cool mountain slopes; coastal areas with mild winters and heavy rains | coniferous evergreen trees (redwood, cedar, hemlock, pine) | bears, elk, wolves, mountain lions |
| Temperate Deciduous Forest | cold to mild winters, long growing season, warm summers, high rainfall | many species; deciduous hardwood trees dominate (elm, maple, oak) | raccoons, squirrels, deer, many different birds |
| Chaparral | rainy, mild winters; hot, dry summers | low shrubs with small leaves (scrub oak, manzanita) | mule deer, coyotes, many lizards and birds |
| Desert | extremely dry; little or no rainfall | cacti, euphorbias, small-leafed or fleshy plants able to withstand heat and drought | small rodents, snakes, lizards, birds |
| Grassland | sparse or intermittent rainfall, temperate | huge, treeless areas covered by grasses | wolves, bison, pronghorns, coyotes, antelope, buffalo |
| Savanna | long dry season | grasses and a few trees (baobab, acacia) | giraffes, lions, zebras, jackals |
| Tropical Rain Forest | high rainfall, warm temperatures year-round | many different species; broadleaf evergreen trees, palms, tree ferns, climbing vines | many different species; bats, birds, lizards, snakes, monkeys |

- **What adaptations can you identify for living things in biomes with extreme climate conditions?**

  Responses include an extra layer of fur, increased fat, hibernation, hard outer covering, ability to store water, tendency to be nocturnal.

- **What features characterize the biome with the highest level of biodiversity?**

  Responses include warm temperatures year round, lots of rainfall.

- **What biome do you live in?**

  Students should identify and discuss features of their biome.

## It's a Wrap

Discuss students' observations and ideas from Activity Sheet 2. Then invite volunteers to share their paragraphs about biomes.

## Home

**Follow Up** Discuss connections to living things that students identified. Were they surprised by their dependence on living things to meet their needs?

**Assignment** Home activity on student page 15.

In your Journal, list the plants and animals that you observe around your home, neighborhood, or community. Use a field guide and Issues and Information section E to help you. Write descriptions for animals and plants that you cannot yet name. Decide which animals or plants are native to your area. Which have special adaptations that enable them to survive in your bioregion?

ACTIVITY SHEET

Name

# DESCRIBE YOUR BIOREGION (part 1)

Biomes are large areas that have the same general climate conditions (extremes of temperature and amount of rainfall), plant life, and animal life. Within biomes there are smaller bioregions that may have special characteristics, such as mountains, rivers, lakes, canyons, and other physical features that can influence plant and animal life. Fill in the chart below to show the characteristics of your local bioregion.

| Feature | Description |
|---|---|
| **CLIMATE**<br>winter<br>summer | **Answers will vary.** |
| **PLANT LIFE**<br>trees<br>shrubs<br>crops | |
| **ANIMAL LIFE**<br>mammals<br>birds<br>insects<br>reptiles | |

ACTIVITY SHEET

Name

# DESCRIBE YOUR BIOREGION (part 2)

| Feature | Description |
|---|---|
| **GEOGRAPHY** altitude, physical features, bodies of water | |
| **ADDITIONAL INFORMATION** | |

1. What was your bioregion like before people lived in it?
   **Answers will vary.**

2. How has your bioregion changed over time?
   **Answers will vary. Students should identify how land has been graded, where dams have been built, and other evidence of imposed changes.**

Explore the Issues: Recognize Biomes 31

# IDENTIFY THREATS TO BIODIVERSITY

Students discover how biodiversity is threatened and what can be done to maintain or increase biodiversity.

## Key Concepts

- People's actions destroy habitats and threaten biodiversity worldwide.
- Natural disasters can cause habitat destruction.
- Habitats can be protected.

## Objectives

After completing this activity students will be able to

- identify ways in which biodiversity is threatened
- identify ways in which biodiversity can be maintained

## Prepare

You may wish to read Issues and Information section C.

## Materials

Activity Sheet 3 for each student

## Setting the Stage

Discuss briefly the questions on student page 17.

- **What human activities threaten biodiversity?**

  Responses include activities that destroy habitats, destroy species, or cause pollution.

- **What can be done to maintain or increase biodiversity?**

  Responses include species protection, habitat preservation, and habitat restoration.

## Vocabulary

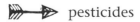 pesticides

## Focus

**1.** Have students study the chart that lists natural and human threats to biodiversity. You may wish to have them read Issues and Information section C. Then have them answer the questions.

- **What are some examples of natural disasters that can threaten biodiversity?**

  Responses include earthquake, flood, volcano, fire.

- **What are some examples of human-caused habitat destruction?**

  Responses include clear-cutting forests, polluting streams, damming rivers.

- **How does habitat destruction threaten biodiversity?**

  Students should be able to explain how species depend on their environment to meet their needs; if the environment is destroyed, they cannot survive.

| Natural Factors | Human-Caused Factors |
|---|---|
| disease | habitat destruction |
| changes in climate | pollution |
| natural disasters | use of pesticides |
| competition between species | exploitation of resources |
|  | introduction of nonnative species |

- **What are some examples of how resources are exploited?**

  Students should recognize that natural resources are used up more quickly than they can be replaced. Examples include cutting down the rain forests, over-fishing oceans and rivers, using up fresh water supplies.

- **How does this threaten biodiversity?**

  Habitats and species are destroyed.

- **How do all of the factors in the chart relate to habitat destruction?**

  Students should give examples of how habitat is affected by each of the threats posed on the chart.

**2.** Have students complete the work on Activity Sheet 3.

## It's a Wrap

Have students share their responses to the questions on Activity Sheet 3. Then have them work together to think of campaign ideas for maintaining or increasing biodiversity. Have them share their posters, public service announcements, or jingles with the group.

## Home

**Follow Up** Have students share their plant and animal lists. They may be able to help each other identify particular species they have observed.

**Assignment** Home activity on student page 18.

 Use your Journal to make a list of activities you have observed at home or at school or in your community that might contribute to habitat destruction and impact biodiversity. Make another list of activities you have observed that enhance habitats and help to maintain biodiversity.

ACTIVITY SHEET

Name

# ACTIONS THAT IMPACT BIODIVERSITY (part 1)

Look at the list of activities below. Decide what impact each one might have on biodiversity. Use your ideas to fill in the chart.

| Feature | Description |
|---|---|
| Plant a garden using a variety of trees, shrubs, and flowering plants. | attract wildlife; increase biodiversity |
| Pour motor oil down a storm drain. | pollute local streams; could kill fish and harm other wildlife; could mean less food for animals that eat fish |
| Clear the brush from a vacant lot. | might remove nesting area for birds or rodents; could eliminate food source (insects) for birds |

34  Environmental ACTION  Habitat  Teacher Resource Guide

ACTIVITY SHEET 3 EXPLORE

Name

# ACTIONS THAT IMPACT BIODIVERSITY (part 2)

| Feature | Description |
|---|---|
| Plant a nonnative groundcover along a highway. | could prevent native plants from spreading; could hold topsoil and moisture to support other plants and animals |
| Use a wood-burning stove to heat your home. | trees cut for fuel destroy forest habitat; wood smoke pollutes the air |
| Hang a birdfeeder in your yard. | attract wildlife; increase biodiversity |

Explore the Issues: Identify Threats to Biodiversity

# MAINTAIN AND INCREASE BIODIVERSITY

Students find out how they can make a difference in maintaining and increasing biodiversity where they live.

### Key Concepts

- Landscaping can be used to help maintain biodiveristy.
- Landscaping can be used to help increase biodiversity.

### Objectives

After completing this activity students will be able to

- identify landscaping practices that attract wildlife
- identify ways to use landscaping to increase biodiversity

### Prepare

You may wish to read background material in Issues and Information section D, as well as information on sustainable landscapes in section G.

Symbols that students may wish to use in their drawings are suggested in section F.

### Materials

Activity Sheet 4 for each student

### Setting the Stage

Discuss briefly the questions on student page 20.

- **How can gardens or landscaping help to maintain or increase biodiversity?**

    Responses include providing species with what they need to survive and using plants that attract wildlife.

- **What landscaping features can help to increase biodiversity?**

    Responses include features that provide food, shelter, and fresh water.

### Vocabulary

 irrigation
landscaping
topsoil

### Focus

**1.** You may wish to have students read about creating sustainable landscapes in Issues and Information section G. Have them look at the list of landscaping practices below and think about how each of the landscaping practices might help to maintain or increase biodiversity. Then have students use the list to answer the questions.

- Plant trees and groundcovers that help conserve water, create shade, and provide wind protection.
- Install a drip irrigation system to conserve water and protect topsoil.
- Avoid use of chemical pesticides and fertilizers.
- Compost garden waste.
- Install a water source, such as a birdbath, fountain, or pond.
- Use plants that produce flowers, berries, seeds, nuts.
- Add ladybugs to the garden.

- **Which of these practices will help provide shelter for birds and other wildlife?**

  Responses include planting trees and groundcovers.

- **Which of these practices will provide water for plants and animals?**

  Responses include installing an irrigation system and a water source.

- **Which of these practices will provide food?**

  Responses include planting plants that produce berries, seeds, and fruit.

- **How do some of these practices help the environment?**

  Responses include helping to curb pollution, prevent erosion, conserve and reuse resources.

- **What other ways can these practices help to maintain or increase biodiversity?**

  Enhancing the habitat will affect biodiversity.

## It's a Wrap

Have students share the ideas they sketched on Activity Sheet 4. Then have them share their ideas about how landscapes can help to maintain biodiversity and how landscapes can help to increase biodiversity.

## Home

**Follow Up** Have students share and compare their lists. What do they think is the greatest threat to biodiversity in the community? What do they think is the most effective measure being taken to maintain or increase biodiversity?

**Assignment** Home activity on student page 21.

Investigate your own back yard or a nearby park. Use your Journal to record your ideas for maintaining or increasing biodiversity in these places. Add to your plant list as you explore the area.

ACTIVITY SHEET 4 EXPLORE

Name

# DESIGN FOR BIODIVERSITY

In the space below, sketch a plan for a park, greenspace, campus, city, or back yard that is designed to maintain or increase biodiversity. Think about what you have learned about landscaping practices and then put your imagination to work.

# Analyze

# LEARN ABOUT COMMUNITY HABITATS

ANALYZE 5

Students learn about local habitat and biodiversity issues from a community resource person.

### Objectives

After completing this activity, students will be able to
- identify features of local habitats
- give examples of how habitats and biodiversity are supported within the community
- describe ways in which habitats and biodiversity within the community can be improved

### Prepare

Make arrangements for a resource person from your community to address the class about habitat and biodiversity issues. Possible speakers include landscaping and gardening professionals, local university or university extension volunteers, community garden group members, habitat and/or wildlife advocates, and land-use representatives.

Activity Sheet 5 may be used as a guide in preparing the presentation. You may also provide a list of the students' questions in advance.

It is not expected that the speaker will be an expert in all areas outlined on Activity Sheet 5. It is important, however, that students understand how the information in the presentation relates to the overall scope of inquiry in this activity. For areas in which the speaker does not have expertise, it would be helpful to suggest where students can go to find needed information.

### Materials

Activity Sheet 5 for each student (See Blackline Masters section.)

### Setting the Stage

Discuss briefly the questions on student page 25.

- **How are habitats and biodiversity supported within the community?**

  Students identify parks and undeveloped land areas in which plants and animals can thrive. Responses also include any known ways in which the community fosters support of habitats and biodiversity (for example, through community awareness programs).

- **How can habitats and biodiversity within the community be improved?**

  Responses include ideas for improving any known conditions that threaten habitats and biodiversity (for example, reduction of available space or nearby use of toxic chemicals); increase green spaces; community projects to clean up streams, protect areas, increase awareness.

### Focus

**1.** Before the speaker comes to class, explain that students will be analyzing habitats and biodiversity in the community. They will examine limitations of the bioregion, how landscaping can foster biodiversity, and how sustainable that landscaping is.

**2.** The speaker will discuss habitats and biodiversity in the community. Students will find out how the resource person functions within the community and about the importance of investigating,

Analyze: Learn About Community Habitats  41

preserving, and improving habitats and sustainable landscaping and gardening within the community. Encourage students to use Activity Sheet 5 to take notes during the presentation.

## It's a Wrap

Have students identify main points for each topic outlined on Activity Sheet 5 and areas that need more investigation.

Have volunteers share their ideas about how the presentation helped them find out more about habitats and biodiversity within the community.

## Home

**Follow Up** Invite students to share and compare their ideas for increasing biodiversity in the community.

**Assignment** Home activity on student page 26.

You have explored biomes and bioregions. Now it is time to investigate an even smaller part of your bioregion. What kinds of habitats are found near your home? What kinds of plants and animals live there? Write a description in your Journal.

ACTIVITY SHEET

Name _____

# NOTES ON COMMUNITY HABITATS AND BIODIVERSITY (part 1)

Use this sheet to record information you learn from the guest speakers who talk to you about landscaping, gardening, and habitat conservation in your community.

Resource Person's Name _____

Title _____

**Local Features** (growing conditions for local zone—temperature, precipitation, soil type, native plants)

**Wildlife Habitats**

(What kinds of wildlife are found in the community?

Where is wildlife found? How do wildlife populations change during the year?

What is being done to help support and encourage wildlife?)

Analyze: Learn About Community Habitats 43

ACTIVITY SHEET 5 ANALYZE

Name

# NOTES ON COMMUNITY HABITATS AND BIODIVERSITY (part 2)

Public Parks and Gardens (purpose of park, types of plants, maintenance)

Conservation Measures (soil, water, energy, animal life, plant life)

Needs of the Community

Plans for the Future

# TOUR THE SCHOOL CAMPUS

Students learn about habitat and biodiversity issues on the school campus from a resource person. They tour the campus and draw a rough sketch of the landscape features.

### Objectives

After completing this activity, students will be able to
- describe habitats and evidence of biodiversity on the school campus
- produce a rough sketch of landscape features on the school campus
- tell how trees and plants are maintained on the school campus

### Prepare

Arrange for a school staff person to address the class about habitat and biodiversity issues, as well as current landscaping and garden practices on the school campus.

Activity Sheet 6 may be used as a guide in preparing the presentation. You may also provide a list of the students' questions in advance. The staff resource person should be familiar enough with the school campus to take students on a tour and answer questions.

### Materials

Activity Sheet 6 for each student

### Setting the Stage

Discuss briefly the questions on student page 28.

- **What kinds of habitats are found on the school campus?**

  Students should describe what they have noticed about the physical features and landscaping on campus.

- **How can landscaping affect biodiversity?**

  Students should identify ways in which landscaping can help to enhance habitats and, in turn, affect biodiversity.

### Focus

**1.** Have students think of questions that they want to ask the resource person and record them in their Journals. The following topics can be used as a guideline:

- how the school grounds relate to features of the local bioregion
- the landscape features of the campus when it was in its natural state
- the school's trees and plants (types and requirements)
- maintenance of the school's trees and plants (water, pesticides/herbicides, soil conditioning, fertilizer, personnel)

Students will tour the campus and find out about the main landscape and garden features. Activity Sheet 6 can be used for taking notes and sketching locations of trees and plants on the school grounds. Details can be added when students return to class. Symbols to use for landscape features can be found in Issues and Information section F.

## It's a Wrap

Have students discuss their work on Activity Sheet 6, identifying components landscape areas on campus. Also provide an opportunity for students to find answers to any questions they still have. Then have volunteers share how the tour helped them find out more about the school's landscape features and needs.

## Home

**Follow Up** Have students share their habitat descriptions.

**Assignment** Home activity on student page 29.

What did the landscape around your home look like before your neighborhood was built? Write a description in your Journal.

ACTIVITY SHEET 6 ANALYZE

Name

# CAMPUS LANDSCAPE FEATURES

In the space below, make a rough sketch of the school campus. Include locations of buildings, parking lots, trees, shrubs, gardens, lawns, and other features that you notice during the campus tour. Include details learned from the guest speaker. Make a key for the symbols you use.

Key—Symbols Used in Sketch

Analyze: Tour the School Campus  47

# PREPARE YOUR AUDIT

Using the sketch from Activity Sheet 6 and their campus tour notes, students will divide the campus into study areas for the audit.

### Objectives

After completing this activity, students will be able to
- identify campus habitats
- use a map to locate landscape features on campus

### Materials

Activity Sheet 7 for each student

### Setting the Stage

Discuss briefly the questions on student page 31.

- **What are the main areas where plants are located on the school campus?**

  Encourage students to think of plant groupings, use areas, and distinct habitats.

- **What features of biodiversity does each region have?**

  Encourage students to identify how plants and animals interact in the areas they have identified.

- **How will you audit campus habitats?**

  Encourage students to share ideas about what they want to find out about species and interdependence on campus and how they will go about collecting data and information.

### Focus

**1.** Have the class use activity sheets and notes to plan a strategy for auditing campus habitats.

**2.** Encourage students to share their ideas about how the campus can be divided up to determine the audit sites they will study.

**3.** Divide the class into Action Groups and assign a study area to each one. Then have the groups meet to organize their audit. Students will record their decisions on Activity Sheet 7.

### It's a Wrap

In their Journals, have students list high-maintenance sites on campus and pinpoint areas with the most wildlife. Encourage students to share their ideas about how these areas can effectively be audited so that important factors are not overlooked.

### Home

**Follow Up** Have students share their ideas about what the landscape looked like before the community was built up.

**Assignment** Home activity on student page 32.

Take a walk around your home. In your Journal, draw a sketch of the buildings, boundaries, and landscaping on your lot.

48 Environmental ACTION  Habitat  Teacher Resource Guide

ACTIVITY SHEET

Name _____

Action Group _____

# CAMPUS AUDIT PLAN (part 1)

Complete the following chart to record your plan for auditing campus habitats.

## Plan for Campus Habitat Audit

| Study Area | Location, Features, General Description, Observations | Action Group (Student Names) | Permission Required/ Accessibility | Audit Due Date |
|---|---|---|---|---|
|  |  |  |  |  |
|  |  |  |  |  |
|  |  |  |  |  |

Analyze: Prepare Your Audit   49

ACTIVITY SHEET

Name _____
Action Group _____

# CAMPUS AUDIT PLAN (part 2)

## Plan for Campus Habitat Audit

| Study Area | Location, Features, General Description, Observations | Action Group (Student Names) | Permission Required/ Accessibility | Audit Due Date |
|---|---|---|---|---|
|  |  |  |  |  |
|  |  |  |  |  |
|  |  |  |  |  |

50 Environmental ACTION Habitat Teacher Resource Guide

# CONDUCT YOUR AUDIT

ANALYZE

Action Groups will inspect their study areas and use their observations to create detailed diagrams.

### Objectives

After completing the activity, students will be able to
- describe the habitat of their assigned study area
- list the plants and animals in their assigned study area

### Materials
- Activity Sheet 8 for each student
- large sheets of drawing paper or butcher paper for each group
- art materials

### Setting the Stage

Discuss briefly the questions on student page 34.

- **What are the features of the habitat in my study area?**

  Students should be able to describe the details of the habitat.

- **What factors affect the biodiversity of my study area?**

  Students should identify positive and negative factors that might affect the living things in their study area. For example, are plants getting trampled by heavy foot traffic? Are birds attracted to trees and shrubs? Is there a constant water source nearby?

### Focus

**1.** Students working in their Action Groups can evaluate their study area and then use their observations to create a large-scale drawing of the site.

**2.** Students can investigate the types of plants and conditions that are found at their site in order to become more familiar with what the area requires and how the area can be enhanced. Field guides and other resources can be used to find out about types of plants and solutions for special problems.

**3.** Have students in Action Groups work together to complete Activity Sheet 8, outlining their strategies for completing the audit.

### It's a Wrap

Have students in their Action Groups review Activity Sheet 8 and the drawings that they made. Have them think of three things they discovered about their study area and three things they need to find out.

Analyze: Conduct Your Audit  51

## Home

**Follow Up** Have students share the sketches they made of the area around their homes. Did any students make surprising observations or discoveries they were not expecting?

**Assignment** Home activity on student page 35.

 Select a natural area that is not actively maintained, such as a back yard, a vacant lot, or a roadway. In your Journal, write your observations about the features of the area.

ACTIVITY SHEET

Action Group

Name

# DETAILS OF STUDY AREA (part 1)

Take a close look your study area. Record details of your observations in the space below. On the back of this sheet, sketch locations of different features in your study area. When you get back to class, use your notes and sketches to help your Action Group create a detailed map of your study area.

**Habitat Conditions in Study Area**

Soil

Water

Sun

Other

Analyze: Conduct Your Audit   53

ACTIVITY SHEET

Action Group _____

Name _____

# DETAILS OF STUDY AREA (part 2)

**Plants in Study Area**

Trees


Shrubs


Ground Cover


Other


Animals Living in Study Area


Special Considerations/Problems

# RESEARCH PLANT SPECIES

Students will research plant species to find out more about their study areas and how they are maintained.

## Objectives

After completing this activity, students will be able to

- identify plants in their research area
- describe what those plants need in order to survive

## Materials

Activity Sheet 9 for each species identified

## Setting the Stage

Discuss briefly the questions on student page 37.

- **In researching plants in your study area, what kinds of information will be useful to you?**

   Responses include size, growth habit, ongoing care, water requirements, hardiness.

- **How will you find the information you need?**

   Responses include asking school maintenance workers, looking in garden books, calling local nurseries or university extension services.

## Focus

Students will use Activity Sheet 9 as a guideline for finding out about the species of plants identified in their study area, including maintenance, care, and habit.

## It's a Wrap

Invite groups to share their findings, as well as strategies for finding information. Can students make some generalizations about which plants require the most upkeep and care? Are some plants more suitable for some areas than another? Based on what they have learned, should any plants be relocated?

## Home

**Follow Up** Encourage students to describe the neglected areas they studied and have them share their observations.

**Assignment** Home activity on student page 38.

Identify three plants that grow around your home, research each of them, and record the information in your Journal.

ACTIVITY SHEET 9 ANALYZE

Name

# DATA SHEET FOR LANDSCAPE PLANTS

Use the following chart to collect information about the plants in your assigned area. Use a separate sheet for each different species of plant.

| Study Area | Plant # |
|---|---|
| Species Name | |
| Approximate Number of Plants | |
| Preferences/Requirements<br>• Climate<br>• Soil<br>• Water<br>• Other | |
| Distinguishing Features | |
| Maintenance Considerations<br>• Watering<br>• Ongoing Care<br>• Pesticides/Herbicides<br>• Other | |
| Information Sources | |

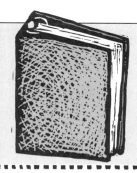

# SUMMARIZE FINDINGS

Students will review audit results, evaluate how habitats are maintained, and describe how biodiversity is supported and write a summary of their findings.

### Objectives

After completing this activity, students will be able to

- describe habitats in their study areas
- describe how biodiversity is encouraged in their study areas

### Materials

Activity Sheet 10

### Setting the Stage

Discuss briefly the questions on student page 40.

- **How are habitats maintained at your study area?**

  Responses should include long- and short-term maintenance needs.

- **How is biodiversity supported at your study area?**

  Responses should describe ways in which species are attracted and encouraged.

- **What problems are associated with the habitat at your study area?**

  Students should describe specific problems, such as high traffic, pollution from parking-lot run-off, bad drainage, and similar challenges.

### Focus

**1.** Meet as a group to compile findings from all of the Action Groups. Each group will give a report about its observations, including the following points:

- landscaping in your group's study area
- landscaping practices that enhance the habitat or encourage biodiversity
- problems such as bad drainage, high traffic, reliance on pesticides and herbicides, labor-intensive maintenance, high water consumption, poor soil
- any other observations about how biodiversity is affected at the site

Students will record information from the class discussion on Activity Sheet 10.

**2.** Students will meet in their Action Groups to discuss and compare the landscaping practices that enhance campus habitats and encourage biodiversity and practices that limit campus habitats and threaten biodiversity. One member of the group should take notes, listing the pros and cons in two columns. For example, on the plus side, a flowering groundcover might prevent erosion, conserve water, and attract birds, butterflies, and insects. On the other hand, the playing field might require lots of water and chemical treatments and high maintenance that discourage wildlife.

**3.** When discussions are completed, group members should work together to write a one-page summary of conclusions that were reached and present the summary to the class.

## It's a Wrap

Discuss campus landscaping practices, pinpointing how habitats can be improved to encourage biodiversity. Then have students write a paragraph in their Journals telling how their group's summary addressed those topics.

## Home

**Follow Up** Have students share information about the plants they researched. Did any students choose the same plant or a member of the same plant family? How could the plants students selected be included in campus study areas?

**Assignment** Home activity on student page 41.

What could you do around your home to improve the surrounding landscape and encourage biodiversity? Write your ideas in your Journal.

ACTIVITY SHEET

Name
Action Group

# CAMPUS HABITATS (part 1)

Combine the audit results from all Action Groups on the chart below.

## Maintaining Campus Habitats and Encouraging Biodiversity

| Study Area | Habitat Features | Maintenance Requirements | How Biodiversity Is Supported | Problems/Observations |
|---|---|---|---|---|
| | | | | |
| | | | | |
| | | | | |

Analyze: Summarize Findings

ACTIVITY SHEET

# CAMPUS HABITATS (part 2)
## Maintaining Campus Habitats and Encouraging Biodiversity

Name
Action Group

| Study Area | Habitat Features | Maintenance Requirements | How Biodiversity Is Supported | Problems/Observations |
|---|---|---|---|---|
|  |  |  |  |  |
|  |  |  |  |  |
|  |  |  |  |  |
|  |  |  |  |  |

# ACT LOCALLY

Students can use the following outreach activities to apply their learning, share information, and strengthen their ties to the community.

1. Organize a native plant sale at your school.

2. Organize a contest or raffle around the theme of preserving habitat and encouraging wildlife. Award conservation-related items, such as books and T-shirts, as prizes.

3. Take a tour of a local garden or nature center. Let them know about your audit and find out how they might be able to assist your efforts to enhance habitats on campus.

# Consider Options

# BRAINSTORM LANDSCAPING IDEAS

Students decide how they can protect or enhance habitats found on the school campus and investigate landscaping practices that will maintain or increase the biodiversity in those habitats.

## Objectives

After completing this activity, students will be able to

- identify landscaping practices that will enhance habitats and increase biodiversity
- evaluate landscaping practices in terms of costs and benefits, responsibility, and long-term goals

## Materials

Activity Sheet 11 for each student

## Setting the Stage

Discuss briefly the questions on student page 47.

- **What landscaping practices can you suggest to increase biodiversity on your school campus?**

  Responses will vary, but evidence of increased biodiversity should be demonstrated.

- **What are the costs and benefits of your suggested landscaping practices?**

  Monetary and nonmonetary costs should be considered.

- **How will protecting or enhancing campus habitats affect costs and benefits of landscaping practices?**

  Students should be able to evaluate their ideas in terms of long- and short-term goals.

## Focus

**1.** Have students review the summaries from the previous activity to pinpoint problems associated with current landscaping practices. List the problems on the chalkboard. Have each Action Group tackle a different problem.

**2.** Action Groups get together and brainstorm solutions to each of the problems they have been assigned. One member of each group can take notes. Students should think about how landscaping practices can be changed to increase biodiversity and enhance habitats. What are the options? Are there better or more efficient ways to protect habitats and increase biodiversity? Students can record their ideas on Activity Sheet 11.

**3.** Once each group has finished brainstorming a list of options, students should go back and briefly evaluate the ideas to get rid of any that simply will not work. Then have them work individually or with a partner to complete Activity Sheet 11.

As students conduct their research, have them keep in mind the following questions for each landscaping idea:

- **What current landscaping practice prompted the need for a change?**
- **What are benefits of this plan in terms of increasing biodiversity and maintaining habitats?**
- **Are any pieces of equipment and/or materials needed? If so, how will we get them?**

- How much time is needed to put the plan into action?
- What is the estimated cost of the plan (including labor) and how can the items be paid for? Can they be donated or loaned?
- Who is responsible for making the plan happen, including maintenance and repair?

## It's a Wrap

Discuss what change each student would make at his or her study site if only one change could be made. How would the change affect biodiversity? How would it improve the habitat?

## Home

**Follow Up** Have students share their ideas for improving the landscape surrounding their home and encouraging biodiversity.

**Assignment** Home Activity on student page 48.

 Identify three landscaping practices at or near your home that help to increase biodiversity. Identify three landscaping practices that could help to protect the habitat at or near your home.

ACTIVITY SHEET 11

Name

Action Group

# LANDSCAPING OPTIONS

Use the following chart to help you organize your ideas while brainstorming landscaping options.

**Problem:**

|  | Options | Research |
|---|---|---|
| **Landscaping Practices that will Increase Biodiversity** |  | Who will research?<br><br><br>Information Sources |
| **Landscaping Practices that will Preserve or Restore Habitat** |  | Who will research?<br><br><br>Information Sources |

Consider Options: Brainstorm Landscaping Ideas 67

# WEIGH THE COSTS AND BENEFITS

Students use the information they have gathered to evaluate the costs and benefits of the landscaping practices they are proposing to adopt.

### Objectives

After completing this activity, students will be able to

- evaluate costs and benefits of landscaping practices
- consider long- and short-term effects of landscaping practices

### Materials

Activity Sheet 12 for each student

### Setting the Stage

Discuss briefly the questions on student page 50.

- **Can you measure or put a value on nonmonetary costs and benefits? Explain your answer.**

  Students should explain how enhancing habitats and increasing biodiversity can beautify the campus and help to maintain or create a pleasing, healthy environment.

- **Why is it important to consider both long- and short-term effects before choosing a landscaping plan?**

  Students should explain how the landscaping plan is likely to evolve over time as the plants become established and care requirements change. Also, students should consider the impact of attracting new species of animals to the habitat. Will the habitat be able to support an influx over time?

### Focus

Students working in their Action Groups will do a cost-benefit study of the options they have chosen. The following questions can help them as they evaluate their ideas:

- **Who will be affected by the change? How?**
- **If the option calls for changing people's habits, how will you get them to cooperate? Is this a cost or a benefit?**
- **What are the nonmonetary benefits of this option?**
- **What are the nonmonetary costs of this option?**
- **Are there any long-term costs or benefits?**

Students can record their ideas on Activity Sheet 12.

### It's a Wrap

Reiterate the importance of considering monetary and nonmonetary costs, as well as short- and long-term effects of landscaping ideas aimed at enhancing habitats and increasing biodiversity. Discuss any ideas that seemed good at the outset but did not hold up once they were evaluated with these considerations in mind.

## Home

**Follow Up** Have students share their ideas for increasing biodiversity and protecting habitats near their homes.

**Assignment** Home activity on students page 51.

 Would any of the alternative landscaping ideas that you learned about be appropriate for your home? Write a description of how you might use one of these ideas at your home.

ACTIVITY SHEET

Name

Action Group

# ASSESS COSTS AND BENEFITS

Use the following chart to evaluate the costs and benefits of each landscaping problem and the options for solving it.

**Problem:**

|  | Costs | Benefits |
|---|---|---|
| **Option 1** | Monetary:<br><br>Nonmonetary: | Monetary:<br><br>Nonmonetary: |
| **Option 2** | Monetary:<br><br>Nonmonetary: | Monetary:<br><br>Nonmonetary: |
| **Option 3** | Monetary:<br><br>Nonmonetary: | Monetary:<br><br>Nonmonetary: |

# MAKE RECOMMENDATIONS

Students working in Action Groups give final consideration to all the landscaping options they have been exploring. After weighing the costs and benefits of each, they will select the best ones and develop a finished proposal for presenting the recommendations to the class.

## Objectives

After completing this activity, students will able to

- describe the criteria they used for determining which landscape practices are the best
- describe why the best solution might not be the most cost-effective solution

## Materials

- Activity Sheet 13
- oak tag, butcher paper, or poster board
- art materials

## Setting the Stage

Discuss briefly the questions on student page 53.

- **How can you decide the best landscaping strategies to implement?**

  Students should describe the factors they should consider as they make their decision.

- **Are the lowest-cost options always the best choices? Why or why not?**

  Students should explain how long-term goals need to be taken into account, as well as site use, what kind of shape the site is in to begin with, and how biodiversity is impacted by existing conditions.

## Focus

**1.** Students get together in Action Groups to decide on which solution to implement, taking into account the costs and benefits and the stated goals.

**2.** Students use Activity Sheet 13 to plan their presentation, adding visual aids as necessary to create a convincing report.

## It's a Wrap

Discuss the criteria that the different groups used to arrive at their decisions. What role did the requirements of each site play in how landscaping practices were determined? How much of a factor was cost? Have volunteers share their Journal entries.

## Home

**Follow Up** Have students share their ideas about implementing a campus landscaping idea at home.

**Assignment** Home activity on student page 54.

Think of ways that you can document or record a part of the natural world around your home (photographs, pressed flowers, sketches). Keep this record in your Journal.

ACTIVITY SHEET 19
CONSIDER OPTIONS

Name
Action Group

# LANDSCAPING PROPOSAL

Use the space below to provide specific data about the proposal your group is making to enhance habitats and increase biodiversity on campus. You will want to include diagrams, graphs, charts, and other graphic organizers to pinpoint the benefits of your proposal.

Problem:

Solution:


Implementation of Proposal

   Action Required:

   Cost:

Long-term Maintenance (costs and labor):


Type of Habitat

   Current Landscaping:

   Proposed Change:

   Benefit:

Effect on Biodiversity:


Other Benefits:

# ACT LOCALLY

Students can use the following outreach activities to apply their learning, share information, and strengthen their ties to the community.

1. Create a one- or two-page list of environmentally friendly landscaping and garden resources—the address of an heirloom seed company, a list of plants that attract butterflies and birds, a pamphlet about how to restore a stream habitat—and include it in the school newspaper or distribute it as a flyer to students.

2. Participate in an organized community group activity aimed at protecting or enhancing the environment, such as a hike, fund-raiser, or clean-up activity.

3. Organize a "clean-up the campus" day at your school.

# Take Action

# CHOOSE LANDSCAPING MEASURES

Action Groups will present their recommendations to the class. The class will discuss the various suggestions and reach a consensus about those recommendations they will include in the final proposal to be presented to school officials.

## Objectives

After completing this activity, students will be able to
- evaluate and compare conservation measures
- reach a consensus with classmates

## Prepare

You may wish to make a transparency of Activity Sheet 14 so that it can be displayed on an overhead projector. As students discuss each presentation, totals can be tallied, averaged, and compared on the sheet in order to help students reach a consensus.

## Materials

- One copy of Activity Sheet 14 for each measure being considered (See Blackline Masters.)
- Transparency of Activity Sheet 14 (optional)

## Setting the Stage

Discuss briefly the questions on student page 59.

- **What factors need to be considered when deciding on which landscaping practices to recommend?**

  Responses include low cost, high benefits, team effort, cooperation among those people affected, ease of maintenance, realistic expectations, suitability for setting.

- **What will help make a landscaping recommendation successful?**

  Responses include participation from schoolmates and staff, effective implementation, increased awareness of the need to protect habitats.

## Focus

Students will use Activity Sheet 14 to rate each presentation. Tally the student responses so that there is a class ranking sheet for each proposal. Encourage discussion as the rankings are interpreted and compiled. Suggest that students find ways to consolidate similar ideas into a single measure. Guide students in analyzing the class responses on the ranking sheet and then use them to prioritize suggestions. Students will then be able to reach a consensus on the measures they wish to present to the school committee.

## It's a Wrap

Follow-up the tallying of the ratings with a discussion of the choices students have made. Emphasize the strengths of the ideas before instructing students to write their paragraphs or draw their cartoons about the conservation measures. Invite students to share their paragraphs or cartoons with the group.

## Home

**Follow-Up** Invite students to share ways in which they documented the natural world around them.

**Assignment** Home activity on student page 60.

 Read a magazine or newspaper article about landscaping or garden practices. Write a summary of it in your Journal.

ACTIVITY SHEET

Name _____

# RATING SHEET

Fill in the following rating sheet for each presentation.

Group _____

Plan _____

Costs

Expensive •   •   • Inexpensive

Environmental Benefits

Low •   •   • High

Impact on Habitat and Biodiversity

Low •   •   • High

Difficulty of Implementing

Low •   •   • High

Cooperation Incentives

Low •   •   • High

Effectiveness of Presentation

Low •   •   • High

Additional Factors to Consider

_____
_____
_____

Priority

Low •   •   • High

Take Action: Choose Landscaping Measures 79

# PREPARE AND PRESENT YOUR PROPOSAL

Students will write a proposal for implementing water conservation measures at school. They will outline costs, benefits, and describe the ways to implement their recommendations. Then they will present it to the school committee.

## Objectives

After completing this activity, students will be able to
- assign tasks to complete a project
- use an outline to prepare a proposal

## Prepare

Discuss the school committee for whom students will be preparing the proposal. Tell them who the members are and describe what you know about the process they will use to evaluate the proposal. You may want to set a date for meeting with the school committee that will be receiving the proposal package.

## Materials

- Activity Sheet 15
- art materials

## Setting the Stage

Discuss briefly the questions on student page 62.

- **What factors had the most serious impact on habitat and biodiversity of the site you audited?**

  Responses should include natural and human threats, location considerations, and other factors.

- **How will the landscaping ideas you are proposing provide both short- and long-term solutions?**

  Students should be able to describe what they expect to have happen as the landscaping matures.

- **What were the most important reasons for choosing these measures?**

  Students should be able to defend their decision-making process.

## Focus

Lead students in a brainstorming session to explore ways of organizing and presenting their proposal.

- **What will be the most effective plan for organizing the presentation? Should you begin with the problem, provide the solution, and then outline the costs and savings? Would another order be more persuasive?**

  Students may debate different organization plans, all of which may be reasonable. Students should consider their audience (the committee) and decide how they can best capture its attention and gain its approval.

- **What is the most important idea you want to emphasize? The importance of increasing biodiversity or enhancing the habitat? The savings? The ease and low cost of implementing the idea?**

  Students will realize that there are many beneficial ideas included in their proposal, but guide them to understanding that a persuasive proposal will be more effective if it is well researched, documented, and presented.

- How can you use charts, graphs, tables, and diagrams to illustrate and promote your ideas?

  Students should suggest ways of displaying information using graphic organizers. They might create a chart showing the kinds of insects, butterflies, and birds they expect to attract; seasonal views of the trees and shrubs they propose planting to show how the site will change during the year.

- **What tone will be the most persuasive?**

  Students should consider what approach will be both diplomatic and credible to the committee members. For example, write the following two statements on the board and guide students in examining which is the most effective and why.

- *The parking strip between the south lot and the music building is a real eyesore.*

  *The south-facing wall of the music building is an idea spot for a climbing vine that will not only provide a habitat for birds but can also help to conserve energy in the building.*

  Invite students to suggest additions or revisions to the outline on the student page. Then have students divide up the tasks and responsibilities and proceed with preparing their proposal.

## It's a Wrap

When students have completed their proposal, allow time for them to reflect on their work and to revise parts of it as necessary before delivering it or presenting it to the committee for consideration.

**Reminder** At this time you may wish to administer the Content Quiz and Student Survey. Your students may already have taken these as pre-tests. They may now be given as post-tests. In addition, the Student Self-Evaluation Form can now be used to help students assess their own progress. These forms are all included in the Blackline Masters/Assessment Tools section of your Resource Guide. Answers to the Content Quiz are on page 16.

## Home

**Follow-Up** Have volunteers share article summaries with the rest of the group.

**Assignment** Home activity on student page 63.

Write a letter about habitat and biodiversity issues to a city official, state or congressional representative, or even to the President. Describe what you like about his or her preservation or enhancement efforts, or make specific suggestions about how habitat management policies can be improved. Put a copy of your letter in your Journal.

ACTIVITY SHEET 15 TAKE ACTION

Name _____

 # PROPOSAL CHECKLIST (part 1)

Use this checklist to plan and monitor tasks that may need to be done in order to complete your proposal. Make a note of who is responsible for completing each task, when each task should be completed, materials needed, and so on. Add to the list as needed.

| TASKS | NOTES |
|---|---|
| **1. TITLE**<br>☐ Cover illustration<br>☐ Proposal statement | |
| **2. WRITE THE INTRODUCTORY PARAGRAPH.**<br>☐ Explain the project.<br>☐ Briefly describe audit findings. | |
| **3. WRITE RECOMMENDATIONS.**<br>☐ Outline each plan.<br>☐ Highlight the benefits.<br>☐ Specify the costs.<br>☐ Suggest a step-by-step plan for implementation.<br>☐ Include ideas for motivating student body, increasing awareness, and encouraging participation.<br>☐ Outline long-range maintenance requirements, costs, planning.<br>☐ Pinpoint projected savings. | Continue your recommendations on next page. |

82  Environmental ACTION  Habitat  Teacher Resource Guide

Name

 **PROPOSAL CHECKLIST (part 2)**

ACTIVITY SHEET

| TASKS | NOTES |
|---|---|
| (CONTINUED) | |
| 4. PRESENT RESEARCH FINDINGS.<br>☐ Prepare graphs.<br>☐ Design tables or charts.<br>☐ Prepare illustrations, photographs, or other art work. | |
| 5. WRITE CLOSING STATEMENT.<br>☐ Describe parts of the plan that are already underway and explain where to go from here. | |

Take Action: Prepare and Present Your Proposal   83

# TRACK RESPONSE TO PROPOSAL

Students will follow up on their landscaping proposal as it is implemented on campus. They will work to increase awareness of the importance of enhancing habitats in order to maintain or increase biodiversity.

## Objectives

After completing this activity, students will be able to

- summarize landscaping practices approved by the school committee and track its implementation
- survey and assess biodiversity and habitat awareness

## Prepare

Review the school committee's response to the landscaping practices students proposed and monitor measures taken. Establish a way to track success and to continue advocacy of water conservation. Activity Sheet 16 can be used as a guide.

## Materials

Activity Sheet 16

## Setting the Stage

Discuss briefly the questions on student page 65.

- **How can you find out what effect your landscaping ideas are having?**

  Responses may include observing wildlife, evaluating plant growth, determining whether problems were solved by the landscaping practices that were recommended and implemented.

- **How can you assess the level of awareness and participation in enhancing habitats and increasing biodiversity?**

  Responses may include interviewing students, observing behavior changes, monitoring the effectiveness of the landscaping practices on campus.

## Focus

Discuss the effects students observe as they see their landscaping ideas implemented. Encourage them to evaluate the impact their work is having on changing people's awareness of habitats and biodiversity.

- **How can other students and staff in your school be motivated to continue to support habitat enhancement?**

  Encourage students to brainstorm methods to keep biodiversity issues before the student body. Examples might include updates in the student newspaper on bird sightings on campus, progress of campus habitat improvements, and awareness of threats to habitats worldwide.

- **How can student participation be increased?**

  Challenge students to think of ways to continue promoting the importance of habitat enhancement and biodiversity protection. Ideas might include establishing a school conservation or bird-watching club, organizing a committee to produce regular awareness posters about threatened habitats worldwide, declaring a quarterly conservation day at school, creating a forum in the school newspaper where all students can contribute ideas or experiences concerning biodiversity and landscaping practices.

## It's a Wrap

Discuss responses to It's a Wrap questions and the ongoing success of students' landscaping plan.

- **What are the most surprising benefits?**

    Possible responses include the amount of savings, the degree of support shown by students and faculty, the enthusiasm of the school committee.

- **What would make the plans more effective?**

    Now that students are seeing the effects of their plan, they should have additional ideas to improve or expand it.

    Challenge students to list more ideas for increasing student awareness of habitat and biodiversity issues.

## Home

**Follow-Up** Encourage students to share any responses they have received from local, state, or national officials.

**Assignment** Home activity on the student page 66.

What can you do? Besides all of the activities that you and your classmates have planned, think of three ideas for action that promote the preservation of habitat and biodiversity, either at home or in your community. Record your ideas in your Journal.

ACTIVITY SHEET

Name

# TRACKING SHEET (part 1)

Use this tracking sheet to summarize and monitor the results of your proposal and to assess students' awareness of habitat and biodiversity issues.

**Proposal Summary**

**Implementation Report**

Month 1 Evidence

Month 2 Evidence

Month 3 Evidence

86 Environmental ACTION   Habitat   Teacher Resource Guide

ACTIVITY SHEET 16 Take Action

Name

# TRACKING SHEET (part 2)

**Recommendations for Increasing Implementation**

**Results**

| | Enhanced Habitat | Increased Biodiversity |
|---|---|---|
| Month 1 | | |
| Month 2 | | |
| Month 3 | | |

**Participation Rating**

Low • • • High

**Suggestions for Increasing Participation**

# Appendices

# Section A
# BIODIVERSITY

Biodiversity is the term used to describe the incredible variety of life on earth—the tremendous array of plants, animals, and ecosystems. It is almost impossible to imagine earth without this rich abundance and variety of life, yet biodiversity is a natural resource that is seriously threatened today. Biodiversity is as much in need of protection as our air, land, and water.

Biodiversity exists at three different levels: ecosystem diversity, species diversity, and genetic diversity. Learning about these different levels will help you understand the full importance of biodiversity.

## Ecosystem Diversity

An ecosystem is a community of living organisms in a particular environment and the nonliving things with which it interacts. A desert, for instance, is an ecosystem that includes organisms such as cacti, lizards, and birds and nonliving parts such as sandy soil, rocks, and sunlight. Ecosystem diversity is used to refer to the various large categories of ecosystems called biomes, such as a desert or rain forest, as well as smaller ecosystems, such as a riverbank or the north side of a mountain. The organisms in an ecosystem depend on each other and on the nonliving parts for survival. (See Section B for more information on ecosystems and biomes.)

## Species Diversity

Species diversity is often what is meant when the term biodiversity is used. Species diversity is the remarkable variety of species of living things—plants, animals, fungi, and microorganisms. It is not known how many different species there are on earth. Over 1.4 million have been identified, and millions have yet to be determined. In the 1980s scientists estimated there were about 5 million species in total. But as they have done more studies, especially in tropical forests where species diversity is enormous, their estimates have gone up and up—to over 30 million. In fact, some research suggests there may be 30 million species of insects alone.

## Genetic Diversity

Each individual organism has thousands of genes that determine its characteristics—its color, height, weight, shape, resistance to disease, and so forth. A common group of genes, called the gene pool, exists for each species. Individual members of a species develop with different genes drawn from the gene pool, which results in genetic diversity. Because of this genetic diversity, organisms can evolve and adapt to changes in their environment, including developing the ability to survive diseases.

Wild plants and animals have more varied characteristics,—that is, more genetic diversity—than cultivated ones. Individual plants of a cultivated plant crop, for example, all have the same genes, and none may have the characteristic to fight a particular disease. If a disease strikes the crop, as occurred in the corn blight in the U.S. in 1970, the entire crop may be lost. Many of the types of tomatoes we eat today would have been lost except for the introduction into their breeding of the genes of a wild Peruvian tomato resistant to disease. The abundant genetic diversity found in wild organisms is the building block for the continuing evolution of life on earth.

## Why Is Biodiversity Important?

Many people would say that the importance of biodiversity is obvious. We experience the value of biodiversity constantly in our enjoyment of nature—in walking in the park, taking trips to the zoo or a wild area, working in our gardens, playing with our pets, reading books and watching TV shows about wondrous creatures in foreign lands. We marvel at nature's diversity. Some believe that biodiversity should be protected just because it is so wonderful. Some say biodiversity is important for other philosophical or spiritual reasons—for a reverence for life and a moral obligation not to destroy what has been created. But there are many other reasons why biodiversity is important.

Think in terms of the three kinds of biodiversity, for instance. Loss in any of these types of biodiversity disrupts the balance of life. If an ecosystem is destroyed, all the organisms adapted to that ecosystem are likely to be destroyed as well. If a species is lost, all the other species whose lives depend on interactions with it may be lost. If it is what scientists call a "keystone" species, the existence of a whole ecosystem may depend on it. If a gene pool is lost or reduced, the genes that make it possible for species to survive through adaptations to changing environments are lost. The loss of genetic diversity, in fact, is considered by many people today to be the greatest threat to the long-range health of our planet.

Biodiversity is also important for the countless direct benefits it provides to humans. The great variety of earth's plants give us the air we breathe; animals and plants give us the food we eat; and a whole array of organisms and microorganisms cleanse the water we drink, regulate floods, recycle waste, and control pests.

Biodiversity also gives us economic and health benefits. Agriculture depends on the genetic diversity of wild plants to improve cultivated crops. Industries, such as wood products and rubber, depend on raw materials from wild plants. Medicine is particularly dependent on biodiversity. Over 40 percent of prescription drugs in the U.S. have been derived from wild species, and new medical treatments from such sources are constantly being discovered.

Many organisms that were once thought to be "useless"—like the penicillium mold or the tropical bat that pollinates precious Asian fruits—have proven to be very valuable. And that is the further tragedy of today's rapid loss of biodiversity—we are losing species before we have any idea what their importance might be.

## Extinction and the Loss of Biodiversity

Extinction is the natural process of species disappearing from the earth—organisms become extinct and new organisms evolve. Before humans appeared on earth, major extinctions were caused by natural environmental changes and occurred over hundreds of years. Today, people's activities—mainly the destruction of habitats—are causing the extinction of species at a rate some researchers suggest is about 400 times the natural rate. This rapid rate of extinction means an alarming loss of biodiversity; nature cannot evolve new species quickly enough to replace those that are being destroyed.

The rate of species extinction has risen steeply from about 400 per year in 1980 to over 10,000 per year in 1990. Some scientists estimate that we are currently losing as many as 17,500 species each year; others estimate that by the year 2000, if human habitat destruction continues as its present rate, the loss of species could be as high as 50,000 each year—or about 130 species every day.

Some researchers estimate that we could lose more than 25 percent of all species on earth within the next few decades. This massive extinction represents a greater biological loss than any in geologic history, including the disappearance of the dinosaurs and other forms of life 65 million years ago. (See section C for further information on the threats to biodiversity and causes of extinction.)

# Section B
# BIOMES, BIOREGIONS, AND HABITATS

In earlier times, people knew their own regions intimately. They knew when plants flowered or seeds matured, when fruits ripened, where birds nested and where fish returned to spawn. They depended on this knowledge of the land, plants, and animals around them to provide all their needs. Some people today are trying to regain some of this closer relationship to their environment by living more simply on the land, raising their own food, or by buying only foods that are locally grown, such as those available at a farmer's market. But generally we are much less well acquainted with our own regions today. What plants and animals live in your region? Which of these were there before humans arrived? Which have been introduced? What is the annual rainfall? What makes up the soil? When does the first flower bloom in spring?

## Biomes

One way to start the study of the region you live in is by identifying which region you are in according to biologists' classifications of the different regions on earth. The land-based regions on earth are classified into large communities called biomes. Each biome is distinguished by its climate (temperature and rainfall patterns) and its communities of plants and animals. Biologists identify between about six and twelve biomes, depending on the level of detail they are using. Nine biomes typically identified are

- tundra
- taiga
- temperate coniferous forest
- temperate deciduous forest
- chaparral
- desert
- grassland
- savanna
- tropical rain forest

Biomes include areas in different regions of the world. There is tundra, for example, on flat land in the Arctic as well as high on the tops of mountains in many different countries. Each biome has similar types of plants and animals, but individual species vary from area to area.

## Biomes and Biodiversity

The different climates and geographic characteristics of the various ecosystems and biomes in a sense cause biodiversity. When plants and animals evolve and adapt to the characteristics of their particular regions, they become different from species in other regions, and thus biodiversity is increased. Many organisms become so specialized to their regions that they cannot survive anywhere else.

The regions with the greatest biodiversity are the tropical rain forests, wetland environments, and the coral reefs in the ocean. Biologists estimate that perhaps two-thirds or more of all species on earth live in tropical rain forests. Twenty-five acres of tropical forest were found to have as many tree species as there are in all of North America; one tropical tree was found to have 43 different kinds of ants; 19 trees to have 12,000 types of beetles; and one river to have more fish species than all the rivers in the U.S. It is probably the relative warmth and abundance of water in tropical forests, wetlands, and coral reef areas that makes them able to support a greater variety of life forms than other regions. Because these delicate environments teem with life, damage to them causes particularly great losses in biodiversity.

The biodiversity of species is distributed unevenly over the globe. The biomes closest to the equator—the rain forests—have the greatest species diversity. The biomes farthest away—the ice and tundra of the Arctic and Antarctic—have the least species diversity. Although scientists do not fully understand the reasons for this uneven distribution of species, the year-round warmth and moisture in the tropical areas may explain it. Without the cold of winter to contend with, organisms can grow and reproduce in the tropics year-round. Insects, for example, can complete their life cycles in a very short time.

## Bioregions and Habitats

Once you have identified your biome, to further study your region you will want to look at the special characteristics unique to your area and the homes of the living organisms around you. Within any biome are numerous bioregions—areas with a unique set of plants, animals, weather conditions, soil conditions, and geographic features such as hills, mountains, lakes, wetlands, or valleys. Animals and plants that have evolved naturally in a particular bioregion are called native plants or animals.

Within any bioregion are more specialized areas that are homes for each organism. These homes are called habitats. The characteristics of a habitat include such features as special soil conditions, light conditions, temperatures, quality and availability of air or water, and the presence of other organisms. The habitat for a bird would be where it has the food, water, shelter, and nesting site it needs. The habitat for a particular shrub might be in the shade where it is very cool and moist—perhaps on the north side but not the south side of a building. A butterfly's habitat might require the presence of a specific plant for its nectar. The habitat for a worm would be underground in certain types of soil.

Habitats for some plants or animals are very specialized; other organisms can live in a variety of habitats. Many organisms—particularly plants—provide the habitat for other organisms. An oak tree often provides the habitat for many different organisms—lichens, mosses, birds, many types of insects, spiders, lizards, rodents; a green lawn might provide the habitat for fewer organisms—a few types of insects, spiders, worms. You will learn a great deal about the different habitats in your area by undertaking the adventure of closely observing and studying the patterns of life in each one.

# Section C
# THREATS TO BIODIVERSITY

Species have always—and will always—die out. They become extinct because of competition with other species and not being able to adapt to environmental changes. Natural events such as shifts in climate or natural disasters (such as major floods or fires) can cause species to be lost. The natural process of extinction usually occurs over many years.

## Effects of Civilization on Biodiversity

As the human population has expanded and required more food and other goods, the rate of species extinction has grown well beyond the natural rate. The primary cause of this rapid loss of biodiversity is the human destruction of habitats. We destroy habitats when we cut down forests for fuel or timber or to create farm or ranch land. We destroy habitats by damming waterways; by draining wetlands; by building roads and cities; and by polluting the air, land, and water. Habitats are being destroyed or damaged in every part of the earth, including the Antarctic and the Arctic. We destroy habitats with the intention of improving or enriching civilization—for products, for food, for homes—but instead we are destroying the very biodiversity that ensures the future of healthy life on earth. (See section A for further information biodiversity and extinction.)

Overexploitation—excessive harvesting or hunting of particular species—is another human activity that threatens biodiversity. The survival of whales has been threatened because of overharvesting for their oil, for example, just as elephants have been overharvested for their ivory tusks, otters for their fur, and various cacti and tropical birds for their beauty. Overharvesting of the trees in a forest causes the destruction of the entire forest habitat.

Another human threat to biodiversity is the introduction of nonnative species of plants or animals to an environment—sometimes on purpose and other times by accident, such as by carrying seeds stuck to clothing or insects aboard cargo ships. The nonnative species may have no enemies in the new environment, and the native species have no defenses against the nonnatives. Thus, the nonnative species may spread quickly, competing with and destroying native species. For example, a third of the native bird species in Hawaii have become extinct due to nonnative snakes, pigs, and other animals brought in on ships from other countries. Similarly, the Oriental kudzu vine, introduced into the southeastern U.S., has become a serious pest, threatening the survival of many other plants.

## The Special Case of Tropical Forests

Because tropical forests are home to such a tremendous variety of species—possibly over 60 percent of all species on earth—the destruction of habitat in tropical forests is the cause of the greatest loss of biodiversity today. Tropical rain forests are being destroyed primarily for fuel; for range land; for land for agriculture; and for the timber to make furniture, paper, and other goods. At least 40 percent of the tropical forests on earth have already been destroyed, and we are destroying something like another 20 hectares (49 acres) every minute. The loss of biodiversity due to human destruction of habitat in the tropical forests is staggering.

# A Lesson From the Past

We often think of ancient civilizations as less exploiting of the environment than our civilization, but archeological research shows that ancient people, too, caused increased species extinction by clear-cutting forests, overharvesting plants and animals, and destroying habitats. We may have a particular lesson to learn from these findings, in fact, for in some cases, the loss of habitat and biodiversity caused by some ancient civilizations may have brought about their own extinction.

One case is that of the extraordinary civilization of the early Maya people in Central America. No one is certain why their culture collapsed around 1600 after thriving for thousands of years, but recent studies have shown that over 80 percent of the forests where they lived had been destroyed. Archaeologists speculate that the loss of the forests may have caused erosion of the soil and thus severely damaged the Mayas' ability to grow corn and other crops important to their survival. Another case is that of the ancient Anasazi people in the southwestern United States. Their over hunting of tigers, mammoths, and bears and overharvesting of vast forest resources may have similarly forced them to leave the sites of their early impressive civilization.

# Section D
# CONSERVING BIODIVERSITY

Scientists are pointing out that conservation of biodiversity needs to be part of how we plan, use, and enhance our natural resources everywhere. People are taking steps to conserve biodiversity by protecting wild habitats, establishing protection laws, and enriching biodiversity in their local areas with parks, gardens, and landscapes. They are also trying new ways of gardening, landscaping, using land, and harvesting wildlife that do as little damage to the natural ecosystems as possible. These new methods are called "sustainable" because they allow the populations and habitats to be sustained for the future.

## Protect Habitat

One of the most important ways to conserve biodiversity is to prevent further loss by protecting wild habitats. Many environmental organizations and governments around the world are working to safeguard natural areas. So far, however, only about three percent of land on earth is protected. And some biomes have much more protected land than others. Notably, there is very little protected land in the rain forest and grassland biomes. Protecting wild habitats is the best way to preserve species, genetic, and ecosystem diversity.

## Establish Protection Laws

The Endangered Species Act in the U.S. is a good example of a law established to protect biodiversity. Species whose survival is threatened are protected under this law because, as the act says, they are of "aesthetic, ecological, educational, historical, recreational, and scientific value to the Nation and its people." International laws, based on treaties and conventions, also protect wildlife. Many people feel strongly about protecting some of the endangered species such as the giant panda, the mountain gorilla, and the California condor, but many of the less sensational, less charming endangered or threatened organisms are equally important to human and environmental survival.

## Practice Sustainable Gardening and Landscaping

People can enhance the biodiversity of their own communities by planting and caring for gardens and landscapes, especially those that imitate the local bioregion. For example, an area of native plants, maintained without pesticides, will become a habitat for many birds and other animals of the region. Similarly, a landscape of plants that mimic the habits and needs of natives—such as drought-tolerant plants in an area with little rainfall—will have the least damaging impact on local resources and provide habitat for other organisms. Gardening not only enriches our environment, it also helps us to understand and appreciate the workings of nature. Issues and Information sections G and H provide more information about practicing sustainable gardening and landscaping.

## Practice Sustainable Harvesting

New sustainable methods of harvesting wildlife are being tried which do not threaten the survival of a species or its ecosystem. Many of these experimental methods do not succeed because of the difficulty of having a thorough knowledge of all the interactions of an intricate ecosystem. But some new methods are succeeding, and it is an important effort to find alternatives to overharvesting. For example, domesticated blue fox are being raised on ranches rather than killing the wild fox for their fur. For another example, selective methods of hunting—taking only a sustainable harvest—have maintained a thriving population of North American deer.

## Take Action at the Local Level: What YOU Can Do

### Start a Project

- Visit a nearby arboretum or botanical garden. Then turn the school campus into an arboretum or botanical garden with signs identifying the variety of trees, shrubs, or other plants. Identify which species are native.
- Use local gardens and parks as the focal point for celebrations and meetings.
- Help to beautify an old cemetery.
- Reclaim a vacant lot and create a community garden where plots can be used on a seasonal or yearly basis.
- Identify the habitats of birds and put up and maintain bird feeders or birdbaths in the areas where they perch and feed. Learn the names of the birds that come to feed or bathe.

### Work with Others

- Get involved with wildlife or gardening organizations.
- Adopt a nearby park and help the park administrators in their beautification efforts, lending a wildlife habitat perspective.
- Adopt a stream and work with city officials to beautify or maintain it or to educate people about how to keep it clean.
- Participate in wildlife appreciation activities such as birding, photography, and whale watching.
- Volunteer for a wildlife rehabilitation organization or an animal shelter.
- Visit and support local and national public lands such as National Wildlife Refuges, Forests, and Parks.

### Educate Yourself and Others

- Write or talk to elected officials about your interest in biodiversity, habitat protection, and natural resource conservation.
- Find out about local habitat-related issues. Express your concern about such issues by writing elected officials or the newspaper.
- Learn more about endangered species by visiting aquariums, botanical gardens, and zoos.
- Educate yourself about products that come from endangered animals and plants, such as coral, furs, exotic shells, cacti, orchids, reptile skins, ivory, tropical birds, and things made of mahogany. Avoid buying those products unless you know they are from a responsible, ecologically safe source.
- Find out about fish and wildlife governmental agencies or programs that protect endangered or threatened species or habitats.
- Teach someone else about the importance of genetic diversity, endangered species, and the preservation of habitats.

### Be a Plant Planter

- Work with the local municipality to organize tree plantings around town. Planting deciduous trees on the southern side of buildings and evergreens on the northern side helps to moderate temperatures in both summer and winter.
- Plant trees, shrubs, or flowers that will attract birds, butterflies, and other animals.
- Design and plant a special garden for small children, elderly people, or people with dis-

abilities. For example, plants that are scented or have an interesting texture could be planted for blind people to enjoy.
- Plant a garden to preserve and demonstrate genetic diversity by using traditional varieties of flowers, fruits, and vegetables and maintaining a seed storage system.
- Plant colorful, low-water maintenance flowering plants to help beautify traffic islands and other public spaces such as community centers, hospitals, and libraries.
- Plant native plants, especially native wildflowers. These plants require little maintenance and water.
- Instead of fences, plant rows of shrubs or trees.

**Be Wild!**
- Work with others to leave some areas—large or small—wild and uncultivated as habitat for insects, reptiles, birds, and other native species.
- Join with others to preserve an important natural area such as a wetland, a grove of ancient trees, or a habitat for an endangered species.

# Section E
# PLANTS

To choose plants that are right for your area and for your purpose, you will want to consult books, magazines, local gardening organizations or agricultural extension offices, nurseries, gardening friends, or other experts. The U.S. Department of Agriculture and other organizations divide areas of the country into different hardiness zones or climate zones. Reference books and nurseries categorize plants by these zones as well as by how much sun or shade and how much water they need. The plants for your garden need to be ones that are hardy in your climate and in your chosen spot.

There will be many plants to choose from for your area. A good way to sort through the selection is to think of what you want in terms of the plants' sizes, shapes, and growth habits. Other considerations for your plant selections might include some special qualities you want in your plants. Here are some examples of qualities to consider:

- Native plants: You might want native plants not only for their beauty but also because they are adapted to your area, because of their attractions for wildlife, and because they tend to be low-maintenance.
- Plants for color at different times of year: Plants may be chosen for their colors including the color of their flowers, their fruits and berries, their year-round leaves, their leaves in autumn, or their bark.
- Plants that produce edible fruits, leaves, roots, and other parts: Popular food plants include tomatoes, beans, lettuce, radishes, potatoes, and squashes. Fruit trees such as plum, cherry, and apple thrive in some areas as well.
- Ground covers and lawn substitutes: Grass lawns are well-suited for areas that are walked and played on frequently. Some types of grass are more disease or drought-tolerant than others, but all lawns require a good deal of water and maintenance. There are many ground covers other than grass, some of them producing flowers, some growing as flat mats, others as knee-high clusters of plants, and many requiring less water and care.
- Plants that produce special fragrances: Many plants are noted for their fragrances—some sweet, some spicy, some resembling other odors such as coffee or lemons. Plants generally are most fragrant in sunny, humid weather. Fragrant plants include lavender, sweet pea, daphne, honeysuckle, gardenia, lilac, mock orange, jasmine, peony, and, of course, roses.
- Plants that attract wildlife: Plants grown without pesticides can be used to attract animals by providing shelter or food. Butterflies, for instance, need food for the caterpillars (such as snapdragons, nasturtiums, roses) and nectar for the adults (such as columbines, Shasta daisies, purple coneflowers). Hummingbirds need flowers with nectar (such as hollyhocks, sages, fuchsias). Seed-bearing plants left in the garden to develop seeds will attract seed-eating birds such as bushtits and chickadees. A list of commonly grown plants that attract wildlife is given on page 101.

**Trees**
Alder
Birch
Bottlebrush
Dogwood
European Mountain Ash
Hawthorn
Mulberry
Oak
Sycamore

**Shrubs**
Arbutus
Barberry
Cotoneaster
Fuschsia
Holly
Mahonia
Pyracantha

**Vines**
Honeysuckle
Wisteria
Bougainvillea

**Low-growing Shrubs or Groundcover**
Calluna
Creeping Rosemary
Erica
Iceplant
Sage
Thyme

# Section F
# SYMBOLS TO USE IN DRAWINGS OF LANDSCAPE PLANS

## Top View

SHRUB

DECIDUOUS OR SHADE TREE

EVERGREEN TREE

GROUNDCOVER

FLOWERS

1 inch = 15 feet

## Side View

EVERGREEN

SHADE

ORNAMENTAL

SHRUB

1 inch = 30 feet

# Section G
# PRINCIPLES OF SUSTAINABLE GARDENING AND LANDSCAPING

The overall principle of sustainable gardening and landscaping is to meet current human needs while protecting and enriching local ecosystems and biodiversity for the future. Sustainable landscapes provide peace, beauty, food, and shelter; through sustainable gardens and landscapes we, in turn, can nurture the earth. We can promote human health, a healthy environment, and help to conserve our most precious resource—biodiversity.

The principles of sustainable gardening and landscaping include using minimal resources and using them efficiently; growing plants that are adapted to the local bioregion and that promote genetic diversity; avoiding chemical additives; and in all ways possible minimizing the impact on natural ecosystems. The methods aim to mimic the local bioregion by being in harmony with local geography, soil conditions, climates, plants, and animals. They nurture and even enhance the natural processes of plant growth, pest management, and soil productivity.

More specifically, the principles of sustainable gardening and landscaping are

- Use minimum amounts of water.
- Use minimum amounts of energy (such as power tools or electricity for water pumps).
- Eliminate or minimize use of chemical fertilizers.
- Eliminate or minimize use of chemical pesticides; use nontoxic methods of pest management instead, including encouraging beneficial animals.
- Eliminate or minimize use of equipment that causes air or noise pollution.
- Recycle garden waste.
- Maximize use of nontoxic, recycled, locally produced garden materials.
- Grow plants appropriate or native to the bioregion (considering soil type and climate).
- To the extent possible, use seeds that promote genetic diversity (nonhybrids and multiple varieties).
- Grow plants and create environments that attract and nurture wildlife.
- Protect topsoil by using appropriate drainage and watering systems.

You will find details about applying these principles in section H. If you do decide to plant a garden or do some landscaping, these are some of the many opportunities and rewards that await you:

- Growing, nurturing, and taking responsibility for living organisms.
- Learning first-hand about the seasons and how all things are interdependent.
- Working with the earth; getting your hands dirty and having fun!
- Growing plants without chemicals.
- Growing your own food.
- Discovering how good "fresh from the garden" food can taste!

## Section H
# TIPS FOR PLANNING AND MAINTAINING A SUSTAINABLE GARDEN

Here are just a few tips for growing a sustainable garden or landscape. You can find much more information—and information specific to your area—from local experts, books, and other resources.

Healthy top soil, rich in beneficial microorganisms, is vital to any garden. Learn about your soil type, the nutrients it contains or needs, and how well water is absorbed by it. The type of soil you have has a strong influence on how often you need to water:

- Sandy—Sandy soil consists of coarse particles and dries out quickly (too quickly for many plants).
- Clay—Clay soil consists of very dense particles, absorbs water very slowly, and stays wet a long time (too long for many plants).
- Loam—Loam is the best garden mixture. It consists of a mix of particles of different sizes that absorb water easily yet dry out slowly. Gardeners add amendments to sandy and clay soils to make them more like loam. They also till or dig the soil to lighten and condition it.

Here are some important tips for maintaining healthy soil:

- Compost garden and kitchen waste and use it to amend and mulch garden soil—composting is sometimes called the foundation of sustainable gardening. Compost not only recycles waste, it provides the best source of enrichment for the soil. It supplies beneficial earthworms, fungi, and other microorganisms, as well as a loamy mix of particles to condition garden soil. Organic compost can be purchased if you do not have your own compost. (See Composting later in this section.)
- Fertilize with organic material (compost, manure)—this saves money, time, and prevents toxic runoff.
- Add natural nutrients to soil as needed—most important are the addition of nitrogen (from green manure or compost) and rock phosphorous.
- Mulch around plants—Mulching means covering the soil with organic material such as leaves, bark, wood chips, hay, and compost. Mulching reduces erosion and the loss of moisture from the soil and keeps weeds from growing. It also protects the soil for the activity of beneficial worms and bacteria and keeps the soil and plant roots cooler in summer and warmer in winter.
- Grow cover crops—Cover crops will enrich the soil in food or other gardens that are replanted each year. Cover crops are quick-growing and have natural bacteria that are able to convert nitrogen in the air into usable nitrogen in the soil. They are grown over the winter and plowed under in the spring. They add organic matter to the soil and provide habitat for beneficial insects, as well as providing nitrogen. Popular cover crops are alfalfa, fava beans, and red clover.
- Rotate crops—Changing from year to year which plants are grown on a piece of land will improve the soil and thereby help keep pests and weeds under control. Although this is particularly important for farmers of larger areas, it will benefit a small garden as well.

### Selecting and Placing Plants

Some plants grow best in warm, dry environments; others thrive in cool, moist conditions. Some grow best in sandy soil, others in clay.

Choose plants that are appropriate to your soil and climate conditions and that suit your purposes. (See sections E, G, and I for more information on plant choices and types of plants.) In addition:

- If practical, plant a mixture of native and nonnative or edible plants to enhance sustainability and attract wildlife.
- Although you will start with young, small plants or seeds, space plants according to their mature sizes. Plants that are uncrowded are more resistant to pests and diseases. But do not plant them too far apart—plants are good companions for each other in various ways, including providing shade, keeping soil moist, and preventing erosion.
- Plant at the right time of year for your chosen plants—some plants do best when planted in spring, others in fall.

## Watering

Sustainable gardens and landscapes use minimum amounts of water by employing planting designs and watering systems that promote efficiency. Some techniques are:

- Grouping—group plants with similar water requirements to make watering efficient.
- Shading—use low-water, larger shrubs or trees to shade other plants. Plants in the shade at least part of the day require less water.
- Planting low-water plants—select plants, shrubs, ground covers, and trees that require less water. This is called *xeriscaping*.
- Use water-saving irrigation systems—efficient systems are drip irrigation and soil soakers. Sprinklers are less efficient because they cause water loss through runoff and evaporation, but where they are needed (for lawns, for instance), use low-output sprinklers with automatic timers.

## Using Energy

Sustainable gardening means minimizing the use of all resources, including energy. To minimize energy:

- Use less water—a lot of energy is used to pump water; saving water saves energy.
- Limit use of gas-powered tools—use hand tools instead!
- Plant trees for shade—deciduous trees planted to shade a building will reduce the need for energy for cooling systems in summer.
- Eliminate or minimize use of chemical fertilizers and pesticides—these chemicals are made from petrochemicals, a nonrenewable source of energy.

## Managing Pests

Synthetic chemical pesticides can have damaging effects on the environment and on human health. In addition, they kill not only the unwanted insects and other animals but also the beneficial ones. The sustainable alternative most in use today is Integrated Pest Management (IPM). This method uses a combination of least-toxic ways of controlling pests, including

- Physical controls and barriers—There are many physical methods including, for example, hand-picking insects or surrounding the garden with ashes or copper strips as a barrier to snails. Laying traps such as rolls of newspaper to attract earwigs is another type of physical control.
- Cultivation techniques—Keeping plants healthy will ensure fewer pest problems, and keeping gardens clean, well-pruned, and free of weeds will reduce habitat for unwanted animals. Dealing with pests quickly will also reduce the chances of serious pest invasions.
- Biological controls—Introducing a pest's predator is an effective technique—for example, putting ladybugs in the garden to eat aphids.
- Companion planting—Some plants attract certain pests while others repel them. Combining the right plants as companions can be an important pest control. For example, the carrot fly attacks carrots but will not lay its eggs on leeks. The leek moth and onion fly attack leeks but are repelled

by carrots. Carrots and leeks are good companions. Many herbs and flowers are also good companions for vegetables.
- Encouraging beneficial animals—Birds, toads, snakes, bats, and other animals are natural insect controllers. Fresh water, bushes for shelter, and food plants will attract birds, for example, which in turn will eat unwanted insects. A Baltimore oriole can eat 17 hairy caterpillars a minute!
- Using safe products—Many products are now available, such as insecticidal soap and herbal repellents that are not poisonous to humans or the environment but are effective against pests. Gardeners can make their own sprays, too, from natural repellents such as garlic, onions, and peppers.

## Choosing Garden Materials

To have as little negative impact on the environment as possible, try to use garden materials that are recycled and nontoxic. For example, old railroad ties make great borders for raised garden beds. When buying new materials, hunt for those that are made from renewable resources, that are produced locally, and that can be recycled.

## Choosing Seeds to Promote Biodiversity

Hybrid seeds, which are most of the seeds available in nurseries and supermarkets, have been bred to produce plants that will grow quickly and be healthy if given the right fertilizers. However, these plants usually will not produce usable seeds. Their seeds either will not grow or will produce plants with unpredictable and unwanted characteristics. To promote genetic diversity, some gardeners today are using "heirloom seeds" instead, nonhybrid seeds that produce plants whose seeds can be used for future generations. As generation after generation of nonhybrids are grown, they add variety to the gene pool. Gardeners are also finding that these heirloom strains are often more disease-resistant and better tasting than the hybrids. (See Resources section for sources of heirloom seeds.) If you grow plants from heirloom seeds, you can promote genetic diversity—and save some money while doing it—if you harvest your own seeds for use in following years. Allow some plants to go to seed and collect the most following years. Allow some plants to go to seed and collect the most healthy, mature seeds from several different plants. Dry the seeds completely and store them (in paper bags or glass jars) in a cool, dry place with a constant temperature.

## Composting

Composting is basic to sustainable gardening. It is nature's way of recycling—decomposing organic material such as yard clippings, leaves, and kitchen scraps into nutrient-rich organic fertilizer. By recycling such elements as phosphorous, calcium, nitrogen, carbon, and microorganisms, composting helps to maintain the balance of ecosystems. Composting can be done using any of several different processes to encourage decomposition by earthworms, bacteria, fungi, and other organisms. One of the easiest and most popular methods is to build a simple compost pile and add red earthworms as the decomposers.

# Section I
# GARDEN AND LANDSCAPE DESIGN

People grow gardens and create landscapes for many reasons. Perhaps the most basic reason is our desire to connect with the earth, to establish a closer relationship with the living organisms that sustain us. We grow gardens for food, for flowers, to attract wildlife. We create landscapes for contemplation, recreation, as artistic expression, as memorials to special people or events. There are as many different reasons for gardening and landscaping as there are garden and landscape designs. Throughout history and throughout the world, people have created gardens and landscapes of incredible beauty and variety.

## History of Gardens and Landscapes

Historians believe that cultivated gardens were first created in the ancient Persian and Egyptian empires as cool places for rest from the heat of the desert. Their water gardens, using pools and water lilies, are among the oldest forms of gardening. The Romans and, later, cultures in the Middle Ages built gardens inside homes or walls as sanctuaries from the outer world. In the Renaissance, gardens filled with statues became elaborate symbols of wealth and status. The classical French gardens, such as the extensive gardens at the Palace of Versailles, were further developments of the Renaissance styles. These French gardens and landscapes, with their Greek and Roman statues and their rigid formality and control, seem to emphasize people's dominance over nature.

In contrast, Asian gardens developed over the centuries to use winding pathways, spots for rest and contemplation, flowing water, and man-made elements such as bridges or statuary that blend into the landscape. The plantings evoke a tranquil atmosphere. These gardens seem to reflect a balance between people's activities and nature.

For years, much of the landscaping in the United States was patterned after the European styles, especially a style that developed in Great Britain using green lawns and flower borders. This style requires a temperate, wet climate or—if that is not the climate—supplemental water, fertilizer, and probably pesticides. To avoid the need for these extra resources in the many bioregions in the U.S. that are not temperate and wet, people have been creating regional styles. They have been developing gardens that aim to balance the natural surroundings with the controlled, cultivated areas. These landscapes are more appropriate to local climates, soils, and plant types. Chaparral gardens in the Southwest and California, luscious shady gardens in the South, flower-filled cottage-type gardens in the Northeast, and cactus gardens in desert areas are a few examples of popular, regional styles.

## Three Basic Garden Designs

Designing gardens and landscapes is like painting or building—the designs are limited only by the materials available, the environment, and your imagination. Gardeners might choose a traditional style, or create their own style, combine different styles, or combine different functions for the garden with different styles. Any garden may be "organic" if it is established and maintained without chemical pesticides and fertilizers. Garden designs may be classified into three basic types: plant gardens, food gardens, and habitat gardens. Of course, a garden is often more than one of these types at the same time.

**Plant Gardens** Many plant gardens are grown for their artistic beauty—for a special combination of colors or forms, for their smells, for variety in textures to see and touch, or for flowers to cut and bring inside. Other plant gardens are designed either to hide or to highlight a building or other structure. Still others are for borders along pathways or to create a quiet resting place. A plant garden may contain only one type of plant—such plant gardens are designed either to hide or to highlight a building or other structure. Still others are for borders along pathways or to create a quiet resting place. A plant garden may contain only one type of plant—such as a rose garden or rhododendron garden—or it may contain many different species.

A basic tip from landscape designers for creating a pleasing plant garden is to keep the design simple by limiting the number of different plants, clustering plants of similar types together, or choosing plants within a particular color range. Some of the traditional plant garden designs are:

*Flower gardens*—groupings of flowering plants, including those that live only one season (annuals) and those that last two or more seasons (biennials and perennials). Trees and shrubs are often included in flower garden designs; colors and forms are grouped for contrast and harmony.

*Asian gardens*—tranquil gardens with soft colors, harmonious groupings of leafy plants (often bamboo or Japanese maples), small bridges, flowing water, and curving pathways. Asian gardens may be shady or in the sun.

*English landscapes and gardens*—green lawns and borders of flowers and hedges. British cottage-style gardens include mixtures of perennial flowering plants and shrubs (often including roses as well as many other flowers). The color range of these gardens is sometimes limited to create a more balanced look.

*Herb gardens*—plants grown for their use as medicines or for flavoring foods—such as oregano for Italian tomato sauces, chamomile for stomach-soothing teas, and lavender for fragrances that delight and calm the mind. Herbs are also grown for the color of their leaves—many are striking silver-grays to blue-greens. Traditional herb gardens often feature sundials, beehives, birdbaths, or figurines as focal points.

*Rock gardens*—natural-looking groupings of rocks with flowering plants, cascading plants, and small shrubs planted among them. Plants may be tucked into cracks or small spaces between rocks. The placement of rocks allows the gardener to include shade-loving plants in cooler spots shaded by the rocks and sun-loving plants in spots where they will be exposed to the sun and to heat from the rocks.

*Water gardens*—formal or informal pools of water planted with water-loving plants (such as water lilies) and often including a surrounding area of moist ground planted with water-tolerant plants (irises and cattails, for example). Styles vary widely, but water gardens must have plants that will produce oxygen to keep the water clear and support any fish or turtles that may be introduced.

**Food Gardens** Many people today are growing vegetable gardens. Some grow food in order to have organic produce (grown without chemical fertilizers or pesticides), some to have better tasting and fresher produce, and many for the pure joy of gardening and growing their own vegetables.

**Habitat Gardens** These gardens are havens for wildlife, designed to attract and meet the needs of birds, beneficial insects, and other living things by providing shelter, space, food, and water. Native plants are common in habitat gardens because they have evolved in interdependence with the animals in the local ecosystems. Habitat gardens might include plants that provide leaves, fruits, seeds, or nectar as food for animals. Areas with rock crevices or leaf litter or weeds are favored by butterflies and other insects. Diversity of plant species and arrangements is important in habitat gardens in order to attract a variety of animals. Habitat gardeners often include additional features such as birdhouses, bird feeders, butterfly houses, bat boxes, tree limbs or snags, and ponds or birdbaths.

## Considerations When Planning a Garden or Landscape

For any type of garden or landscape, gardeners must consider not only their vision and desired purpose for the garden but also the limitations of the climate, the space, and the materials available. Books, magazines, gardening organizations, and services such as agricultural extension offices abound as resources for in-depth advice about garden planning. The following list is a starting point to give you some idea of the things to consider when planning a garden or landscape:

### Practical Considerations

- Function of garden (for food, for flowers, to attract wildlife, etc., or a combination of functions)
- Space available
- Existing structures (buildings, walkways, paths that will impact the garden design)
- Water availability and location of source
- Impact on people or ecosystem (allergy considerations, for example, or shading or disturbing neighboring areas)
- Features for special needs such as for disabled people, children, elderly people, or pets
- Cost considerations: cost of building materials (stones, bricks, wood); cost of plants and supplies
- Need for maintenance and cost of maintenance

### Ecological Considerations

- Climate (temperature, wind, rainfall, etc.)
- Exposure to sun or shade
- Soil type and condition
- Water flow (drainage patterns)
- Topography—the shape of the land
- Plants native or adapted to local bioregion
- Variety of plants and seeds that promote genetic and species diversity
- Impact on wildlife and environment

### Artistic Considerations

- Overall design (formal or informal, traditional or nontraditional, one style or a blend of styles)
- Color
- Smell
- Plant proportions
- Use of nonplant elements (benches, walkways, rocks)

# Section J
# ORGANIC GARDENING

An organic garden can bring delight in producing the healthiest food possible, beautiful flowers, and glorious herbs. Because organic gardens use materials from living things such as organic soil amendments, pest controls, and fertilizers, they enhance biodiversity without harming the environment.

You may decide to grow an organic garden on your school grounds, in your back yard, or on a vacant lot. Here are eight basic steps for creating and maintaining an organic garden and a guide to creating a seed bank. Be sure to consult books and local experts for more information specific to your area.

### Step 1: Divide tasks and responsibilities.

Decide who will be responsible for the various tasks of creating, maintaining, and harvesting your garden. If you are growing a garden at school, much of the growth will take place during the summer. Be sure you can make arrangements to regularly maintain the garden during the summer months.

### Step 2: Choose the garden area and develop a detailed plan.

Consider the size of the garden plot, the plants desired, the availability of sunlight and water, and access pathways in the garden. Consider what you want to do with the fruits, vegetables, herbs, or flowers you harvest and choose the right plants for your purposes. For instance, will you want to take the produce home to eat, sell it at a produce stand, donate it to the community, or perhaps share it with your school's food preparers for use in the school cafeteria? To plan for harvests throughout as much of the year as possible, think about your growing seasons and how long different plants take to mature. Plan your planting times and spacings accordingly.

### Step 3: Research and obtain garden supplies.

Here is a suggested list of supplies:
- Square-nosed spade
- Hay or pitch fork
- Small trowels
- A few feet of chicken wire
- String or twine
- Garden stakes (or sticks)
- Irrigation/watering materials
- Organic soil conditioner (compost is best)
- Planting soil (for germinating seeds)
- Paper cups or transplanting trays
- Seeds and/or plants

### Step 4: Prepare the soil.

One popular method of preparing garden soil is called "double digging." This method enriches the soil on the top level and breaks up the lower soil so that roots can grow well. Double digging results in garden beds of raised mounds of soil ready for planting.

a. Use the stakes and twine to mark the areas for your garden beds. A good size is 3 feet by 6 feet.

b. If the area is covered with grass, remove it (compost or use it elsewhere). Try to save any topsoil by shaking it off the roots.

c. Dig a trench one spade deep and set the soil to the side. Use the spade or pitchfork and dig down another spade's depth and loosen the soil. Add amendments to the soil in the lower level.

d. Replace the topsoil and dig the next trench right next to it. Move on down the line until the entire bed is "fluffed" or loosened. Be sure to pull out all the old roots, weeds, and rocks as you work the soil.

e. Add a thin layer of soil amendments (compost) across the top of the bed and work it lightly into soil.
f. Repeat for all the beds in the garden.
g. Don't step on the beds after they have been dug or you will pack them down again!

**Step 5: Plant plants or seeds.**

Determine the appropriate planting times and methods for your plants. To efficiently cover the surface of your garden bed, stagger the spacing of your plants (or seeds) rather than planting them in rows. Use chicken wire to guide your planting of vegetables such as lettuce and other greens, radishes, and carrots. Plant one plant in each chicken-wire hole (or every other hole for larger plants). Seeds can be germinated in the ground or in plastic pots, transplanting trays, or paper cups filled with planting soil.

Moisten the soil before planting seeds or transplanting seedlings. Try to transplant in cool weather or in the evening. When transplanting, dig the hole in the garden soil several inches wider and deeper than the rootmass, rough up the edges of the rootmass, place the plant in the hole, and fill in the hole with a mix of soil and soil amendments. Water the plants in well and don't let them dry out while they are getting established.

**Step 6: Water regularly and manage weeds and pests.**

Determine your plants' needs for water and establish a watering system and schedule. Check regularly for over- or under-watering. If possible, set up an irrigation system with timers to ensure regular watering. Remove weeds and deal with pests as soon as they are found—this will eliminate more work in the long run. Mulch around plants to reduce weeds and keep the soil moist.

**Step 7: Harvest crops and collect seeds.**

Because different crops ripen at different times, harvesting is an ongoing process. It's pretty obvious when crops such as tomatoes are ripe and ready to be harvested, but you may have to consult books or experts to determine the right time for harvesting some of your plants. If you allow some plants to go to seed, you can collect their seeds for planting in your garden next year or for storing in a seed bank (see below).

**Step 8: Prepare the garden for winter.**

Rake up all the old dead plant material and put it in the compost pile.

Spread fresh compost and other mulch (fall leaves are great) on top of the garden beds and dig in lightly. Bacteria and other organisms can then decompose the compost further, and the eggs, larvae, and pupae of unwanted insects will be exposed to winter cold and birds. You might also want to prepare the soil of a small area to be ready for planting early spring crops such as peas and spinach. Preparing the bed before winter avoids having to dig in the wet soil of early spring.

## Creating a Seed Bank

If you have grown nonhybrid plants (see "Choosing Seeds to Promote Biodiversity" in section H), you may want to harvest seeds and save them in a seed bank for your own use later and to contribute to genetic diversity.

a. Allow some plants to go to seed. Collect seeds from several different plants, choosing the seeds that look healthy and mature.
b. Dry the seeds completely.
c. Research the different storage needs of specific seeds. Most seeds will last well for one year in paper bags. For longer storage, most need to be sealed well in glass jars. Beans and peas, however, require air and must always be stored in paper bags.
d. Store the seeds in a cool, dry place with a constant temperature. Both heat and humidity must be avoided, but most seeds will not be harmed by temperatures below freezing.
e. Establish a simple and accurate method for organizing the seeds in your seed bank. Label containers with contents and dates and maintain a record or catalog of all the seeds.

# GLOSSARY

**biodiversity** The variety of living things on earth.

**biome** A major regional or global biotic community, such as a grassland or desert, characterized by the main forms of plant life and the prevailing climate.

**bioregion** A smaller part of a biome characterized by plant life and climate but also taking into account geography, waterways, and other special features.

**chaparral** A biome characterized by hot, dry summers and cool, moist winters and dominated by a dense growth of mostly small-leaved evergreen shrubs.

**climate** The meteorological conditions, including temperature, precipitation, and wind, that characteristically prevail in a particular region.

**coniferous** Needle-leaved or scale-leaved, chiefly evergreen, cone-bearing trees or shrubs such as pines, spruces, and firs.

**deciduous** Falling off or shed at a specific season or stage of growth; shedding or losing foliage at the end of the growing season.

**ecosystem** An ecological community together with its environment, functioning as a unit.

**extinction** The fact or condition of ceasing to exist.

**genetic** Of or relating to genetics or genes.

**habitat** The area or type of environment in which an organism or ecological community normally lives or is found.

**interdependence** Mutual dependence.

**irrigation** To supply dry land with water by means of ditches, pipes, or streams; water artificially; to make fertile or vital as if by watering.

**landscape (v.)** To adorn or improve a section of ground by contouring and by planting flowers, shrubs, or trees.

**pesticide** A chemical used to kill pests, especially insects.

**precipitation** Any form of water, such as rain, snow, sleet, or hail, that falls to the earth's surface.

**savanna** A flat grassland of tropical or subtropical regions.

**species** A basic category of related organisms capable of interbreeding; an organism belonging to such a category.

**taiga** A subarctic, evergreen coniferous forest of northern Eurasia located just south of the tundra and dominated by firs and spruces.

**temperate** Characterized by moderate temperatures, weather, or climate; neither hot nor cold.

**topsoil** The upper part of the soil.

**tropical** Of, occurring in, or characteristic of the Tropics; hot and humid.

**tundra** A treeless area between the icecap and the tree line of Arctic regions, having a permanently frozen subsoil and supporting low-growing vegetation such as lichens, mosses, and stunted shrubs.

# Teacher Resources

# ORGANIZATIONS

## Organizations Providing Educational Materials and Information

**America the Beautiful Fund**

219 Shoreham Building
Washington, DC 20005
(202) 638-1649
Fax: (202) 638-1687
WWW: <http://www.americashmall.com>
WWW: <http://www.charities.org>

The fund grants free vegetable, flower, and herb seeds to schools to improve neighborhoods, create new parks and community gardens, start new environmental education programs, and grow food for the hungry. Schools can apply for a grant for up to 500 pounds of bulk vegetable seeds. Send SASE to Department E2.

**Bio-Integral Resource Center**

P.O. Box 7414
Berkeley, CA 94707
(510) 524-2567
Fax: (510) 524-1758
E-mail: <birc@igc.apc.org>

BIRC is a nonprofit organization that provides practical information on the least-toxic methods for managing pests based on the principles of Integrated Pest Management, or IPM.

Products & Services: Offers publications on Integrated Pest Management; pests of the home and garden; pests of people and pets; agricultural and community-wide pests; and biological control and least-toxic pesticides.

**Ethical Science Education Coalition**

167 Milk Street #423
Boston, MA 02109-4315
(617) 367-9143
(860) 875-1808
Fax: (617) 523-7925
E-mail: <info@ma.neavs.com>

A national organization of educators, students, and concerned citizens who want to ensure that a student's right to a quality education need not be compromised by his/her ethical belief regarding the use of animals in the classroom.

Products & Services: Assists schools in formulating and implementing dissection policies and disseminates information regarding alternative non-animal teaching tools. Frog Fact Sheet; Turtle Fact Sheet; Choice Dissection Policy Sheet; and Beyond Dissection—Innovative Teaching Tools for Biology Education, a catalog of alternatives to dissection.

**The National Arbor Day Foundation**

100 Arbor Avenue
Nebraska City, NE 68410
(402) 474-5655
Fax: (402) 474-0820
WWW: <http://www.arborday.org>

A nonprofit education organization. Members are dedicated to a mission of tree planting, conservation, and environmental stewardship.

Products & Services: Educational materials include booklets, brochures, and flyers on conservation of trees, multimedia curriculum kits, guides to growing your own trees, and tree identification booklets.

### National Coalition Against the Misuse of Pesticides

701 E Street SE, Suite 200
Washington, DC 20003
(202) 543-5450
Fax: (202) 543-4791
E-mail: <ncamp@igc.apc.org>

NCAMP is a broad coalition of health, environmental, labor, farm, consumer, church groups, and individuals who share common concerns about the potential hazards associated with pesticides. The coalition seeks to focus public attention on the very serious pesticide poisoning problem and promotes reduced pesticide exposure through alternative pest management strategies. The coalition also advocates public policies that better protect the public from pesticide exposure.

Products & Services: Safety at Home: A Guide to the Hazards of Lawn and Garden Pesticides and Safer Ways to Mange Pests; packets on Fertilizers, Pesticides on Food, Mercury in Paint, School IPM, and Pesticides in Schools.

### National Pesticide Telecommunications Network

(800) 858-7378, M–F, 8:30 AM–6:30 PM Pacific Time

Hotline sponsored by Oregon State University. Provides scientific and factual information about pesticides: safety, health effects, and environmental effects. Will provide referrals for lab analysis in the case of human and animal poisonings and information on lawn care, pest control, pesticides, and soil testing.

### National Wildflower Research Center

4801 La Crosse Avenue
Austin, Texas 78739
(512) 292-4200
Fax: (512) 292-4627

A nonprofit education and research organization committed to the preservation and re-establishment of native wildflowers, grasses, shrubs, and trees.

Products & Services: Educational fact sheets and packets available at a minimal charge.

### The Nature Conservancy

1815 North Lynn Street
Arlington, VA 22209
(703) 841-5300
Fax: (703) 841-1283
WWW: <http://www.tnc.org>

The Nature Conservancy sponsors the Adopt-an-Acre Program, which provides a direct way to protect the earth's threatened rainforests. Rescue the Reef is another program that involves protecting the coral reef in the Florida Keys, the Caribbean, and the Pacific. These programs involve individuals and professionals working together to save our nation's diverse and sensitive habitats, as well as to educate the public on the rainforest, coral reefs, and other threatened ecosystems.

Products & Services: a bimonthly magazine; informational posters and fact sheets; Nature Conservancy; and Teacher's Guides for the Adopt programs.

### Northwest Coalition for Alternatives to Pesticides

P.O. Box 1393
Eugene, OR 97440
(541) 344-5044
Fax: (541) 344-6923
E-mail: <ncap@igc.apc.org>
WWW: <http://www.enf.org/~ncap>

NCAP is a nonprofit, grassroots organization that promotes sustainable resource management, prevention of pest problems, use of alternatives to pesticides, and the right to be free from pesticide exposure.

Products & Services: The coalition offers a comprehensive information service on the hazards of pesticides and alternatives to their use; information packets, books; fact sheets and the quarterly *Journal of Pesticide Reform*.

### Rachel Carson Council

8940 Jones Mill Road
Chevy Chase, MD 20815
(301) 652-1877
Fax: (301) 951-7179
E-mail: <rccouncil@aol.com>

This is a resource center offering educational publications on pesticide use and pesticide alternatives.

Products & Services: *Alternative Pest Control for Lawns and Gardens; Least-Toxic Ant Control; Least-Toxic Cockroach Control; Pesticides and Lawns* (a table of the active ingredients in 72 commonly used pesticides); Lists the characteristics and hazards of these lawn and garden chemicals; *The Other Road to Flea Control; Toxic Playgrounds.*

### Rodale Institute

611 Siegfriedale Road
Kutztown, PA 19530-9749
(610) 683-6383
Fax: (610) 683-8548

The core concept of the Institute is "Healthy Soil, Healthy Food, Healthy People®." The Institute works worldwide to achieve a regenerative food system that improves environmental and human health. Their programs attempt to find agricultural solutions to hunger, malnutrition, disease, and soil degradation.

Products & Services: The Institute features a bookstore; a 333-acre experimental farm which offers tours, student field trips, and weekly workshops; Food, The Essence of Life, an interactive, hands-on exhibition; Gardenfest!, an annual celebration of organic gardening and food.

### Seed Savers Exchange

3076 North Winn Road
Decorah, IA 52101
(319) 382-5990
Fax: (319) 382-5872

A nonprofit organization that exists to support genetic diversity by growing and exchanging seeds. Members maintain thousands of heirloom varieties of seeds: those that have been inherited; varieties that have been dropped from seed catalogs; and outstanding foreign varieties. Each year many of the members distribute the seeds to ensure their survival.

Products & Services: Provides information on easy seed saving techniques; has published *Seed To Seed,* a comprehensive reference book.

### The Wilderness Society

900 Seventeenth Street NW
Washington, DC 20006-2596
(202) 833-2300
Fax: (202) 429-3957
E-mail: <twsnw@tws.org>

The mission of the Wilderness Society is to ensure the preservation and proper management of all of the nation's public lands—the National Parks, National Seashores and Recreation Areas, National Forests, National Wildlife Refuge System, and the national resource lands administered by the Bureau of Land Management.

Products & Services: *Saving Our Ancient Forests; The Wild World of Old-Growth Forests: A Teacher's Guide;* plus other publications on the wildlife, the Alaska oil spill, and national forests.

### World Wildlife Fund

1250 24th Street NW
Washington, DC 20037
(202) 293-4800
Fax: (202) 293-9211

Products & Services: Offers pamphlets on the rainforest, endangered species, and Windows on the Wild, environmental education materials focusing on the issues of biodiversity.

# GOVERNMENT AGENCIES

### Bureau of Land Management

Office of Public Affairs, Room 5600
1849 C Street NW
Washington, DC 20240
(202) 452-5125
Fax: (202) 452-5124
WWW: <http://www.blm.gov>

An agency within the Department of Interior, the BLM is responsible for ecosystem management of the nation's public lands (270 million acres) and 570 million acres of Federal mineral resources. Their mission is to sustain the health of the public lands for present and future generations.

Products & Services: Provides a number of brochures on their strategies for managing rivers, streams, wetlands, back country byways, fish and wildlife, cultural resources, and more. The brochures include contact numbers and addresses for feedback from the public.

### Environmental Protection Agency

Office of Pesticide Programs
Waterside Mall
401 M Street SW
Mail Stop 7506C
Washington DC 20460
(703) 305-7090

Responsible for regulating the use of pesticides and registering pesticides in the United States.

Products & Services: Provides a number of brochures and fact sheets on the safe use, storage, and disposal of pesticides. They provide a care packet that includes most of their publications such as *IPM in Schools; Healthy Lawn, Healthy Environment;* and the Protecting Endangered Species from Pesticides poster.

### United States Department of the Interior

U.S. Fish & Wildlife Service
1849 C Street NW
Washington, DC 20240
(202) 208-5634
Publication Unit: (703) 358-1711
WWW: <http://www.fws.gov>

The U.S. Fish and Wildlife Service is the principal agency through which the Federal Government carries out its responsibilities to conserve, protect, and enhance the nation's fish and wildlife and their habitats for the continuing benefit of people. The Service's major responsibilities are for migratory birds, endangered species, certain marine mammals, and freshwater and anadromous fish.

Products & Services: Publications on endangered and threatened species, wetlands, wildlife conservation.

### United States Department of the Interior

National Park Service
1849 C Street NW
Mail Stop 3220
Washington, DC 20240
Public Affairs: (202) 208-6843 (public affairs)
WWW: <http://www.nps.gov>

Almost all of the sites in the National Park System (over 360 sites) offer some type of interpretative or educational program. They may range from publications, video presentations, and guided walks and talks to curriculum-based education programs. To find out more about these programs contact the national park site nearest to you.

# BOOKS AND PAMPHLETS

### Backyard Composting: Your Complete Guide to Recycling Yard Clippings

Harmonious Technologies
P.O. Box 1716
Sebastopol, CA 95473
(707) 823-1999

Everything you need to know about composting and more. Includes a resource list of environmental books, videos, and organizations.
94 pages.
$6.95

### The Backyard Wildlife Habitat™ Information Kit

Craig Tufts
National Wildlife Federation, 1993
1400 Sixteenth Street
Washington, DC 20036-2266
(800) 822-9919
(800) 432-6564

A how-to guide on planting to attract a variety of different wildlife, regardless of the size of your area. 80 pages. Item #79946.
$4.95 plus $3.50 for shipping and handling

### Bugs, Slugs & Other Thugs: Controlling Garden Pests Organically

Rhonda Massingham Hart
Storey Communications, Inc., 1991
Pownel, VT 05261
(800) 441-5700

This is for gardeners who have lost their fair share of the garden to damaging insects and wildlife pests. It is a practical guide to pest control that is safe for the user and the environment. 214 pages.
$12.95

### Butterfly Gardening, Creating Summer Magic in Your Garden

Xerces Society/Smithsonian Institution
Sierra Club Books, 1990
Sierra Club Bookstore
85 2nd Street
San Francisco, CA 94105
(800) 935-1056

Instructions for designing and planting gardens to attract butterflies, with garden design diagrams and a master plant list. Also includes regional listings of the most familiar North American butterflies and moths and their nectar and larval food plants, resources for equipment and gardening/conservation organizations, and more. 192 pages.
$19.00

### Common Sense Pest Control: Least-toxic solutions for your home, garden, pets and community

William Olkowski, Sheila Daar, Helga Olkowski
Taunton Press, 1991
63 South Main Street
P.O. Box 5506
Newtown, CT 06470

A complete reference guide to pest management programs for: indoor and outdoor landscapes; greenhouses; buildings; animals, the human body; and more. 715 pages.
$39.95 plus shipping

### Four-Season Harvest

Eliot Coleman
Chelsea Green Publishing Co., 1992
White River Junction, VT 05001

A how-to guide for harvesting fresh, organic vegetables from your garden all year long. 212 pages.
$19.95

**Greening the Garden, A Guide to Sustainable Growing**

Dan Jason
New Society Publishers, 1991
P.O. Box 189
Gabriola Island, BC V0R 1X0
(800) 567-6772

This book suggests that gardening and growing one's own food makes a personal as well as a political statement. With an eye on practicing bioregionalism, the book includes growing tips, nutritional information, recipes, and more. 208 pages.
$12.95

**The Mulch Book: A Complete Guide for Gardeners**

Stu Campbell
Storey Communications, 1991
105 Schoolhouse Road
Pownal, VT 05261

Details everything one needs to know for mulching. 120 pages.
$9.95

**The Organic Gardener's Handbook of Natural Insect and Disease Control: A Complete Problem-Solving Guide to Keeping Your Garden and Yard Healthy Without Chemicals**

Edited by Barbara W. Ellis and Fern Marshall Bradley
Rodale Garden Books, Rodale Press 1992
(800) 848-4735

Directions on how, when, and where to use preventative methods, insect traps and barriers, biocontrols, homemade remedies, botanical insecticides, and more. Over 350 color photos for identification. 534 pages.
$17.95 (paperback)

**Pest Control in the School Environment: Adopting Integrated Pest Management**

U.S. Environmental Protection Agency, August 1993
Office of Pesticide Programs (H7506C)
401 M Street SW
Washington, DC 20460-0003
(202) 260-7751
(202) 260-2080
Fax: (202) 260-6257

This booklet provides students with a general understanding of IPM principles so that they can make an informed decision about pest control in their schools. 43 pages.

**Rodale's All-New Encyclopedia of Organic Gardening: The Indispensable Resource for Every Gardener**

Edited by Fern Marshall Bradley and Barbara W. Ellis
Rodale Press, 1992
(800) 848-4735

A comprehensive resource book for gardening, pest control, raising food crops, as well as how to maintain perennials, annuals, trees, shrubs, and lawns, all without chemicals. 690 pages.
$29.95

**The Schoolyard Habitats™ Information Kit**

The National Wildlife Federation
8925 Leesburg Pike
Vienna, VA 22184-0001
(800) 477-5560

A how-to guide that outlines the entire process of creating a Schoolyard Habitat—an area set aside on or near school grounds that invites wildlife and provides hands-on learning opportunities for students and teachers alike. Includes a pre-paid application for certification in the program. Item #79948. $14.95 plus $3.50 for shipping and handling

**Solar Gardening**

Leandre Poisson and Gretchen V. Poisson
A Real Goods Independent Living Book, 1994
(800) 762-7325

Provides guidelines for growing year round, from the hottest months to the coldest months. Also includes instruction for increasing the square-foot yield of your garden, extending the growing and harvest seasons for vegetables, and building solar appliances for the garden. 296 pages. Item #80-247.
$24.95

**Worms Eat My Garbage®**

Mary Appelhof
Flower Press, 1997
10332 Shaver Road
Kalamazoo, MI 49024
(616) 327-0108
Fax: (616) 327-7009

A complete how-to for setting up and maintaining a worm composting system. 100 pages.
$10.95

# PRODUCTS AND SERVICES

**A-1 Unique Insect Control**

5504 Sperry Drive
Citrus Heights, CA 95621
(916) 961-7945
Fax: (916) 967-7082

Source for ladybugs, praying mantis egg cases, earthworms, lacewing eggs, trichogramma, and fly parasites for the biological control of destructive pests. First class and priority mail shipments made around the country.

**Bountiful Gardens**

19550 Walker Road
Willits, CA 95490
Catalog free in U.S.

A project of Ecology Action, Bountiful Gardens is a resource for herb, flower, grain, covercrop, and heirloom vegetable seeds; books; a list of beneficial insects; organic gardening supplies.

**D. Landreth Seed Co.**

180-188 West Ostend St.
Baltimore, MD 21230
Catalog $3

This is America's oldest seed house and counts George Washington, Thomas Jefferson, and Joseph Bonaparte among its early customers. Today it supplies a wide variety of seeds for vegetables, flowers, everlastings, fruits, and herbs.

**Gardens Alive!**

5100 Schenley Place
Lawrenceburg, IN 47025
(812) 537-8650
Fax: (812) 537-5108
E-mail: <76375.2160@compuserve.com>

A catalog of organic garden products, featuring a variety of items for enhancing garden growth and controlling a range of garden pests. The free catalog is mailed on a monthly basis; you may want to request mailings only when the products are changed quarterly.

**Harmony Farm Supply**

3244 Hwy. 116 No. H
Sebastopol, CA 95472
(707) 823-9125
Fax: (707) 823-1734
WWW: <http://harmonyfarm.com>

Catalog featuring irrigation systems, fertilizers, beneficial insects, tools, and supplies.

**IPM Laboratories, Inc.**

P.O. Box 300
Locke, NY 13092-0300
(315) 497-2063
Fax: (315) 497-3129
E-mail: <ipmlabs@baldcom.net>

Offers a catalog of biological controls, tools, and supplies.

**J.L. Hudson, Seedsman**

P.O. Box 1058
Redwood City, CA 94064
Catalog $1

J.L. Hudson offers a wide range of rare and unusual varieties of heirloom seeds. He is dedicted to open-pollinated seeds and the preservation of our genetic resources. In addition, his catalog is a fascinating and invaluable resource.

### Molbak's Seattle Garden Center

1600 Pike Place
Seattle, WA 98101
Write for information

The seed selection available by mail order covers more that 125 varieties of herbs, including damassia, a natural slug and snail killer.

### Seeds of Change

P.O. Box 15700
Santa Fe, NM 87506-5700
(800) 95-SEEDS for a free catalog
(505) 438-8080
Fax: (505) 438-7052
E-mail: <seedchange @aol.com>
WWW: <http://www.seedsofchange.com>

The catalog features certified organic seeds suitable for growing just about anywhere in the United States. Seeds have been chosen in part for their nutritional value rather than just their appearance or uniformity. Seeds of Change is dedicated to providing a diverse selection of varieties including traditionals, and heirlooms. The catalog carries nutritious food plants, medicinals, herbs, culinary herbs, and flowers. It also provides informative reading, as well as tools and books for gardeners both young and old.

### Shigo and Trees Associates

P.O. Box 769
Durham, NH 03824
(603) 868-7459
Fax: (603) 868-1045

Offers posters, books, videos, and brochures about trees and tree care.

### Soil Testing Kit

Real Goods
555 Leslie Street
Ukiah, CA 95482-5576
(800) 762-7325
Fax: (707) 468-9486
WWW: <http://www.realgoods.com>

This kit provides all materials necessary for determining pH (30 tests) and concentrations of nitrogen, phosphorus, and potassium (15 tests each). Also included is the *Garden Guide Manual* and *Soil Handbook* with instructions for interpreting test results and fertilizer recommendations. Item #57-132.
$39.95

### Southern Exposure Seed Exchange (SESE)

P.O. Box 170
Earlysville, VA 22936
(804) 973-4703
Fax: (804) 973-8717
WWW: <http://www.southernexposure.com>

The catalog specializes in heirloom seeds and emphasizes varieties adapted for the Mid-Atlantic region, but successfully serves gardeners throughout the U.S. and Canada. The seeds are free of chemical treatment; many are organically grown and have been germination tested. SASE provides information and supplies for saving your own seeds.
$2.00 (catalog updated annually)

### Suppliers of Beneficial Organisms in North America

California EPA
Department of Pesticide Regulation
1020 N Street, Room 161
Sacramento, CA 95814-5604
(916) 324-4100
Fax: (916) 324-4088
E-mail: <chunter@cdpr.ca.gov>

A listing of suppliers of beneficial organisms used for the biological control of pests. Single copies are free. An electronic edition is available at <http://www.cdpr.ca.gov/docs/dprdocs/goodbug/organism/htm>

### Worm-a-Way®

Flowerfield Enterprises
10332 Shaver Road
Kalamazoo, MI 49024

(616) 327-0108
Fax: (616) 327-7009

The kit includes Mary Appelhof's vermicomposting bin, made from recycled plastic, plus her book *Worms Eat My Garbage,* a worm fork, and one pound of redworms.
$76

**Worm's Way**

7850 N. State Road 37
Bloomington, IN 47404
(800) 274-9676
Fax: (812) 876-6478
WWW: <http://www.wormsway.com>

Catalog features fertilizers, pest control methods, growing kits, water gardening, hydroponics, environmental controls, and more.

# Blackline Masters

Name _____

ACTIVITY SHEET

# IMPORTANCE OF BIODIVERSITY (part 1)

Study the illustration and answer the questions that follow.

**1.** What features of this habitat can you identify?

_____

_____

_____

Recognize Biodiversity   127

ACTIVITY SHEET

Name _____

# IMPORTANCE OF BIODIVERSITY (part 2)

**2.** What could happen to the biodiversity in this habitat if a flood wiped out the corn crop?

_____
_____
_____

**3.** Mice are an important source of food for owls. How could the owls be affected by the loss of the corn crop?

_____
_____
_____

**4.** How would the owls be affected by the loss of the hawks?

_____
_____

**5.** What other examples of interdependence can you identify?

_____
_____
_____

**6.** What could you do to increase the biodiversity in this habitat?

_____
_____

**7.** What could happen if this habitat were developed into a shopping mall?

_____
_____
_____

Name _____

ACTIVITY SHEET

# DESCRIBE YOUR BIOREGION (part 1)

Biomes are large areas that have the same general climate conditions (extremes of temperature and amount of rainfall), plant life, and animal life. Within biomes there are smaller bioregions that may have special characteristics, such as mountains, rivers, lakes, canyons, and other physical features that can influence plant and animal life. Fill in the chart below to show the characteristics of your local bioregion.

| Feature | Description |
|---|---|
| **CLIMATE** winter  summer | |
| **PLANT LIFE** trees  shrubs  crops | |
| **ANIMAL LIFE** mammals  birds  insects  reptiles | |

Recognize Biomes

ACTIVITY SHEET

Name _____

# DESCRIBE YOUR BIOREGION (part 2)

| Feature | Description |
|---|---|
| **GEOGRAPHY** <br> altitude <br><br> physical features <br><br> bodies of water | |
| **ADDITIONAL INFORMATION** | |

**1.** What was your bioregion like before people lived in it?
_____

**2.** How has your bioregion changed over time?
_____
_____
_____

ACTIVITY SHEET

Name _____

# ACTIONS THAT IMPACT BIODIVERSITY (part 1)

Look at the list of activities below. Decide what impact each one might have on biodiversity. Use your ideas to fill in the chart.

| Feature | Description |
|---|---|
| Plant a garden using a variety of trees, shrubs, and flowering plants. | |
| Pour motor oil down a storm drain. | |
| Clear the brush from a vacant lot. | |

Identify Threats to Biodiversity   131

ACTIVITY SHEET

Name

# ACTIONS THAT IMPACT BIODIVERSITY (part 2)

| Feature | Description |
|---|---|
| Plant a nonnative groundcover along a highway. | |
| Use a wood-burning stove to heat your home. | |
| Hang a birdfeeder in your yard. | |

ACTIVITY SHEET 4 EXPLORE

Name _____

# DESIGN FOR BIODIVERSITY

In the space below, sketch a plan for a park, greenspace, campus, city, or back yard that is designed to maintain or increase biodiversity. Think about what you have learned about landscaping practices and then put your imagination to work.

ACTIVITY SHEET

Name _____

# NOTES ON COMMUNITY HABITATS AND BIODIVERSITY (part 1)

Use this sheet to record information you learn from the guest speakers who talk to you about landscaping, gardening, and habitat conservation in your community.

Resource Person's Name _____

Title _____

- - - - - - - - - - - - - - - - - - - - - - - - - - - - - - - - - - - - - - - - - - - - -

Local Features (growing conditions for local zone—temperature, precipitation, soil type, native plants)




Wildlife Habitats

(What kinds of wildlife are found in the community?

Where is wildlife found? How do wildlife populations change during the year?

What is being done to help support and encourage wildlife?)

134   Learn About Community Habitats

ACTIVITY SHEET 5 ANALYZE

Name _____

# NOTES ON COMMUNITY HABITATS AND BIODIVERSITY (part 2)

Public Parks and Gardens (purpose of park, types of plants, maintenance)

Conservation Measures (soil, water, energy, animal life, plant life)

Needs of the Community

Plans for the Future

Learn About Community Habitats   135

ACTIVITY SHEET

Name _____

# CAMPUS LANDSCAPE FEATURES

In the space below, make a rough sketch of the school campus. Include locations of buildings, parking lots, trees, shrubs, gardens, lawns, and other features that you notice during the campus tour. Include details learned from the guest speaker. Make a key for the symbols you use.

Key—Symbols Used in Sketch

136  Tour the School Campus

ACTIVITY SHEET

# CAMPUS AUDIT PLAN (part 1)

Name _____  Action Group _____

Complete the following chart to record your plan for auditing campus habitats.

## Plan for Campus Habitat Audit

| Study Area | Location, Features, General Description, Observations | Action Group (Student Names) | Permission Required/ Accessibility | Audit Due Date |
|---|---|---|---|---|
|  |  |  |  |  |
|  |  |  |  |  |
|  |  |  |  |  |
|  |  |  |  |  |

Prepare Your Audit 137

**ACTIVITY SHEET**

# CAMPUS AUDIT PLAN (part 2)

## Plan for Campus Habitat Audit

Name _____    Action Group _____

| Study Area | Location, Features, General Description, Observations | Action Group (Student Names) | Permission Required/ Accessibility | Audit Due Date |
|---|---|---|---|---|
| | | | | |
| | | | | |
| | | | | |

138  Prepare Your Audit

© The Tides Center/E2: Environment & Education

ACTIVITY SHEET

Name

Action Group

# DETAILS OF STUDY AREA (part 1)

Take a close look your study area. Record details of your observations in the space below. On the back of this sheet, sketch locations of different features in your study area. When you get back to class, use your notes and sketches to help your Action Group create a detailed map of your study area.

**Habitat Conditions in Study Area**

Soil


Water


Sun


Other

Conduct Your Audit 139

ACTIVITY SHEET

Name　　　　　　　　　　　　　　Action Group

 # DETAILS OF STUDY AREA (part 2)

**Plants in Study Area**

Trees

Shrubs

Ground Cover

Other

---

Animals Living in Study Area

---

Special Considerations/Problems

ACTIVITY SHEET 9 ANALYZE

Name _____

# DATA SHEET FOR LANDSCAPE PLANTS

Use the following chart to collect information about the plants in your assigned area. Use a separate sheet for each different species of plant.

| Study Area | Plant # |
|---|---|
| Species Name | |
| Approximate Number of Plants | |
| Preferences/Requirements<br>• Climate<br><br>• Soil<br><br>• Water<br><br>• Other | |
| Distinguishing Features | |
| Maintenance Considerations<br>• Watering<br><br>• Ongoing Care<br><br>• Pesticides/Herbicides<br><br>• Other | |
| Information Sources | |

Research Plant Species 141

ACTIVITY SHEET

Name _____

Action Group _____

# CAMPUS HABITATS (part 1)

Combine the audit results from all Action Groups on the chart below.

## Maintaining Campus Habitats and Encouraging Biodiversity

| Study Area | Habitat Features | Maintenance Requirements | How Biodiversity Is Supported | Problems/Observations |
|---|---|---|---|---|
| | | | | |
| | | | | |
| | | | | |

142   Summarize Findings

ACTIVITY SHEET

Name

Action Group

# CAMPUS HABITATS (part 2)

### Maintaining Campus Habitats and Encouraging Biodiversity

| Study Area | Habitat Features | Maintenance Requirements | How Biodiversity Is Supported | Problems/Observations |
|---|---|---|---|---|
| | | | | |
| | | | | |
| | | | | |

Summarize Findings 143

ACTIVITY SHEET 11
CONSIDER OPTIONS

Name _____

Action Group _____

# LANDSCAPING OPTIONS

Use the following chart to help you organize your ideas while brainstorming landscaping options.

**Problem:**

| | Options | Research |
|---|---|---|
| **Landscaping Practices that will Increase Biodiversity** | | Who will research?<br><br>Information Sources |
| **Landscaping Practices that will Preserve or Restore Habitat** | | Who will research?<br><br>Information Sources |

144   Brainstorm Landscaping Ideas

© The Tides Center/E2: Environment & Education

ACTIVITY SHEET 12
CONSIDER OPTIONS

Name
Action Group

# ASSESS COSTS AND BENEFITS

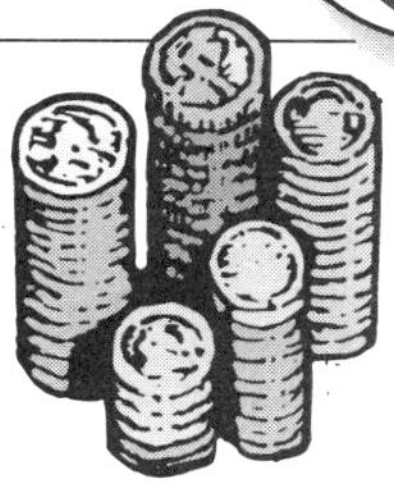

Use the following chart to evaluate the costs and benefits of each landscaping problem and the options for solving it.

**Problem:** _____

|  | Costs | Benefits |
|---|---|---|
| **Option 1** | Monetary:<br><br>Nonmonetary: | Monetary:<br><br>Nonmonetary: |
| **Option 2** | Monetary:<br><br>Nonmonetary: | Monetary:<br><br>Nonmonetary: |
| **Option 3** | Monetary:<br><br>Nonmonetary: | Monetary:<br><br>Nonmonetary: |

Weigh the Costs and Benefits

ACTIVITY SHEET

Name

Action Group

# LANDSCAPING PROPOSAL

Use the space below to provide specific data about the proposal your group is making to enhance habitats and increase biodiversity on campus. You will want to include diagrams, graphs, charts, and other graphic organizers to pinpoint the benefits of your proposal.

Problem:

Solution:

Implementation of Proposal

   Action Required:

   Cost:

Long-term Maintenance (costs and labor):

Type of Habitat

   Current Landscaping:

   Proposed Change:

   Benefit:

Effect on Biodiversity:

Other Benefits:

Name _____

ACTIVITY SHEET

# RATING SHEET

Fill in the following rating sheet for each presentation.

Group _____

Plan _____

**Costs**

Expensive •         •         • Inexpensive

**Environmental Benefits**

Low •         •         • High

**Impact on Habitat and Biodiversity**

Low •         •         • High

**Difficulty of Implementing**

Low •         •         • High

**Cooperation Incentives**

Low •         •         • High

**Effectiveness of Presentation**

Low •         •         • High

**Additional Factors to Consider**

_____
_____
_____

**Priority**

Low •         •         • High

Choose Landscaping Measures   147

ACTIVITY SHEET

Name

# PROPOSAL CHECKLIST (part 1)

Use this checklist to plan and monitor tasks that may need to be done in order to complete your proposal. Make a note of who is responsible for completing each task, when each task should be completed, materials needed, and so on. Add to the list as needed.

| TASKS | NOTES |
|---|---|
| **1. TITLE**<br>☐ Cover illustration<br>☐ Proposal statement | |
| **2. WRITE THE INTRODUCTORY PARAGRAPH.**<br>☐ Explain the project.<br>☐ Briefly describe audit findings. | |
| **3. WRITE RECOMMENDATIONS.**<br>☐ Outline each plan.<br>☐ Highlight the benefits.<br>☐ Specify the costs.<br>☐ Suggest a step-by-step plan for implementation.<br>☐ Include ideas for motivating student body, increasing awareness, and encouraging participation.<br>☐ Outline long-range maintenance requirements, costs, planning.<br>☐ Pinpoint projected savings. | |

Continue your recommendations on next page.

ACTIVITY SHEET

Name

# PROPOSAL CHECKLIST (part 2)

| TASKS | NOTES |
|---|---|
| (CONTINUED) | |
| **4. PRESENT RESEARCH FINDINGS.**<br>☐ Prepare graphs.<br>☐ Design tables or charts.<br>☐ Prepare illustrations, photographs, or other art work. | |
| **5. WRITE CLOSING STATEMENT.**<br>☐ Describe parts of the plan that are already underway and explain where to go from here. | |

Prepare and Present Your Proposal 149

ACTIVITY SHEET

Name _____

# TRACKING SHEET (part 1)

Use this tracking sheet to summarize and monitor the results of your proposal and to assess students' awareness of habitats and biodiversity issues.

**Proposal Summary**

**Implementation Report**

　　Month 1 Evidence

　　Month 2 Evidence

　　Month 3 Evidence

150　Track Response to Proposal

ACTIVITY SHEET

Name _____

# TRACKING SHEET (part 2)

**Recommendations for Increasing Implementation**

---

**Results**

|  | Enhanced Habitat | Increased Biodiversity |
|---|---|---|
| Month 1 |  |  |
| Month 2 |  |  |
| Month 3 |  |  |

**Participation Rating**

Low •          •          • High

**Suggestions for Increasing Participation**

Track Response to Proposal 151

Name _____

# CONTENT QUIZ (part 1)

Circle the correct answer for each question. Some questions may have more than one correct answer.

1. Biodiversity refers to
    a. the way plants are divided on earth
    b. the huge variety of living things on earth
    c. the number of extinct species on earth
    d. the number of two-legged animals on earth

2. Along with air and water, biodiversity is a threatened resource that needs protection.
    a. true      b. false

3. The biome with the most biodiversity is located
    a. near the mountains
    b. near the poles
    c. near the equator
    d. near the oceans

4. Living and nonliving things in an ecosystem interact.
    a. true      b. false

5. Genetic diversity enables living things to
    a. adapt to environmental changes
    b. overcome diseases
    c. evolve over time
    d. all of the above

6. Threats to biodiversity upset the balance of life.
    a. true      b. false

7. An entire ecosystem can depend on one keystone species.
    a. true      b. false

8. The biodiversity of species is distributed evenly throughout the world.
    a. true      b. false

Name _____

# CONTENT QUIZ (part 2)

9. Habitat destruction can be caused by
   a. damming rivers
   b. clear-cutting forests
   c. flash flooding
   d. all of the above

10. Habitat destruction is the greatest threat to biodiversity.
    a. true          b. false

11. Overharvesting of trees can
    a. attract wildlife
    b. destroy the forest habitat
    c. prevent fire
    d. increase biodiversity

12. Protecting habitats will help to
    a. prevent air pollution
    b. increase biodiversity
    c. beautify public spaces
    d. all of the above

Name _____

# STUDENT SURVEY

There are no correct answers to the questions below. Answer "yes" or "no."

　　YES　NO

1. ☐　☐　I know about how habitats are protected.

2. ☐　☐　I know which bioregion I live in and can identify its features.

3. ☐　☐　I encourage my family to protect habitats.

4. ☐　☐　I know that my family enjoys outdoor habitats.

5. ☐　☐　I try to find ways to keep water from becoming polluted.

6. ☐　☐　I think that clean water and clean air are valuable resources.

7. ☐　☐　I know that we do not dump toxic materials down the drains in our home.

8. ☐　☐　I know that we do not spray toxic chemicals on our garden.

9. ☐　☐　My family tries to find alternatives to using materials that might pollute the groundwater.

10. ☐　☐　I know that the best way to encourage biodiversity is to protect habitats.

Name _____  Date _____

Action Group _____

# STUDENT SELF-EVALUATION FORM

Evaluate your contributions to Action Group and class activities.

| Poor | Average | Good | Excellent | |
|------|---------|------|-----------|---|
|      |         |      |           | Actively participated in Action Group/class discussions. |
|      |         |      |           | Demonstrated a clear understanding of issues. |
|      |         |      |           | Took responsibility for research and other tasks. |
|      |         |      |           | Provided suggestions and solutions to problems. |

How could I have improved my participation in this project?

_____

_____

_____

What skills did I improve upon? _____

_____

Which need further improvement? _____

_____

**As a result of participating in this project:**

What new knowledge have you gained? Have you shared knowledge with others at school?

_____

_____

How has your opinion about environmental issues/problems changed? _____

_____

Have you made any changes in your daily life and/or home? _____

_____

_____

Assessment Tools

Action Group _____

Date _____  Time _____

# ACTION GROUP EVALUATION FORM

| Behavioral Functions | Participants' Names | | | | |
|---|---|---|---|---|---|
| | | | | | |
| 1. Initiating discussion | | | | | |
| 2. Sharing information | | | | | |
| 3. Seeking information | | | | | |
| 4. Giving opinions | | | | | |
| 5. Seeking opinions | | | | | |
| 6. Evaluating information | | | | | |
| 7. Clarifying information | | | | | |
| 8. Dramatizing points | | | | | |
| 9. Coordinating tasks | | | | | |
| 10. Finding compromises | | | | | |
| 11. Suggesting procedures | | | | | |
| 12. Recording information | | | | | |
| 13. Finding middle ground | | | | | |
| 14. Relieving tension | | | | | |
| 15. Finding norms | | | | | |
| 16. Withdrawing from group | | | | | |
| 17. Blocking discussion | | | | | |
| 18. Seeking recognition | | | | | |
| 19. Horsing around | | | | | |
| 20. Advocating for group members | | | | | |
| 21. Dominating discussion | | | | | |
| 22. Criticizing others | | | | | |

(Adapted from *Effective Group Discussion,* 4th edition, by John K. Brilhart, Wm. C. Brown Company Publishers, Dubuque, IA)

# PROGRAM EVALUATION FORM (part 1)

Teachers: In an effort to continually upgrade this module, we are asking for constructive criticism and suggestions for improvement. Please complete an evaluation form for each module and return to:

E2: Environment & Education
P.O. Box 20515
Boulder, CO 80308-3515
303/442-3339 Fax: 303/442-6633
email: e2ee@enviroaction.org

Name

School

School address

Grade(s) taught

---

1. Please check the module(s) that you used:
    - [ ] Energy Conservation
    - [ ] Water Conservation
    - [ ] Waste Reduction
    - [ ] Food Choices
    - [ ] Habitat and Biodiversity
    - [ ] Chemicals: Choosing Wisely

2. What did you like most about the module(s)?

3. What did you like least about the module(s)?

4. Please comment on the following aspects of the modules:

    Readability

    Instructions

    Organization

    Graphics and Layout

Assessment Tools 157

Name _____

# PROGRAM EVALUATION FORM (part 2)

Readability
_____
_____

Instructions
_____
_____

Organization
_____
_____
_____

5. Rate how the module affected your students' environmental awareness.

|  | Not at all |  |  |  | Very much so |
|---|---|---|---|---|---|
| Awareness of issues | 1 | 2 | 3 | 4 | 5 |
| Knowledge of issues | 1 | 2 | 3 | 4 | 5 |
| Sense of responsibility | 1 | 2 | 3 | 4 | 5 |
| Sense of ability to effect change | 1 | 2 | 3 | 4 | 5 |

6. Rate the overall involvement of the following groups in conducting and implementing environmental activities at your school.

|  | Not at all |  |  |  | Enthusiastically |
|---|---|---|---|---|---|
| Students | 1 | 2 | 3 | 4 | 5 |
| Maintenance Staff | 1 | 2 | 3 | 4 | 5 |
| School Administrators | 1 | 2 | 3 | 4 | 5 |
| Parents | 1 | 2 | 3 | 4 | 5 |
| Teachers | 1 | 2 | 3 | 4 | 5 |
| Local Community Organizations | 1 | 2 | 3 | 4 | 5 |
| Local Businesses | 1 | 2 | 3 | 4 | 5 |
| Others _____ | 1 | 2 | 3 | 4 | 5 |